Electrophysiologic Testing

in Disorders of the Retina, Optic Nerve, and Visual Pathway

Gerald Allen Fishman, MD

Department of Ophthalmology
University of Illinois at Chicago College of Medicine

Samuel Sokol, PhD

Department of Ophthalmology
Tufts University School of Medicine
New England Medical Center

AMERICAN ACADEMY OF OPHTHALMOLOGY

American Academy of Ophthalmology

655 Beach Street

P.O. Box 7424

San Francisco, CA 94120-7424

Design and Production
Christy Butterfield Design

Library of Congress Cataloging-in-Publication Data

Fishman, Gerald Allen, 1943–
Electrophysiologic testing in disorders of the retina, optic nerve, and visual pathway / Gerald Allen Fishman, Samuel Sokol.
p. cm. — (Ophthalmology monographs ; 2)
Includes bibliographical references.
ISBN 1-56055-004-X
1. Electroretinography. 2. Electrooculography. 3. Visual evoked response. I. Sokol, Samuel. II. Title. III. Series.
[DNLM: 1. Electrooculography. 2. Electroretinography. 3. Evoked Potentials, Visual. 4. Optical Nerve Diseases—diagnosis. 5. Retinal Diseases—diagnosis. WW 270 F537ea]
RE79.E4F57 1990
617.7'307'547—dc20
DNLM/DLC
for Library of Congress 90-449
CIP

Printed in Singapore through Palace Press La Jolla.

Contents

Preface

This monograph is intended to serve as both an introduction and a reference for ophthalmologic residents and practicing ophthalmologists interested in the fundamentals of electrophysiology and its application in various retinal, optic nerve, and visual pathway disorders.

The beginning sections of each chapter include information on techniques of recording and principles of photoreceptor, retinal pigment epithelial cell, or visual pathway physiology as they are reflected in the various components of the electroretinogram, electro-oculogram, and visually evoked cortical potential. The chapters then provide information on electrophysiologic findings in various hereditary and acquired disorders of the retina, optic nerve, and visual pathway.

Although most readers of this monograph are unlikely to ever personally perform the tests described, it is of appreciable advantage to comprehend the fundamental physiologic principles on which the tests are based and to be aware of the circumstances in which they may be of value for diagnosing various diseases or disorders, for better understanding impaired visual function, and for monitoring disease progression.

The authors would be particularly gratified if this monograph stimulated some of our colleagues to pursue investigations in electrophysiology that might ultimately help resolve numerous unanswered questions about various visual disorders.

This monograph is dedicated to the late Alex E. Krill, MD, and Jerome T. Pearlman, MD, in memory of what they contributed to the field of clinical electrophysiology within their all-too-short lifetimes.

Acknowledgments

The authors acknowledge with appreciation Drs Kenneth R. Alexander, Donnell Creel, Francisco de Monasterio, Anne Fulton, Alan Kimura, Robert W. Massof, Anne Moskowitz, Neal S. Peachey, David R. Pepperberg, Harris Ripps, Barry Skarf, and Ken Wright, who reviewed portions of the manuscript during its development. Drs Neal Peachey and Phyllis Bobak made contributions to the sections on oscillatory potentials and the pattern ERG, respectively. Gratitude is extended to Ms Adrienne Adelman and Ms Pat Hornibrook, who typed the manuscript, and Ms Kathy Louden, who edited the manuscript while in development. Dr Sokol expresses his gratitude to Professor Lorrin Riggs, who introduced him to the recording of evoked cortical and retinal potentials in humans.

Research reported in this monograph was supported in part by core grant EY01792 from the National Eye Institute, to the University of Illinois at Chicago Department of Ophthalmology, a center grant from the National Retinitis Pigmentosa Foundation Fighting Blindness (Dr Fishman), research grant EY00926 (Dr Sokol) from the National Eye Institute, National Institutes of Health, and a Research to Prevent Blindness Scientific Investigator Award (Dr Sokol).

CHAPTER 1

The Electroretinogram in Retinal Disease

Gerald Allen Fishman, MD

The full-field light-evoked electroretinogram (ERG) is the record of a diffuse electrical response generated by neural and nonneuronal cells within the retina. The response occurs as the result of light-induced changes in the transretinal movements of ions, principally sodium and potassium, in the extracellular space. This retinal potential can be recorded in all vertebrates and in many invertebrates. It was first identified in recordings from a fish eye in 1865 by the Swedish physiologist Frithiof Holmgren, who initially misinterpreted the waveform as arising from action potentials within the optic nerve. Although Dewar[1] of Scotland recorded this potential in humans as early as 1877, electroretinography did not find widespread clinical application until 1941, when Riggs[2] introduced a contact-lens electrode for human use. In 1945, Karpe[3] reported the results of a study on 64 normal and 87 abnormal human eyes, providing the initial groundwork for clinical investigation.

The ERG represents the combined electrical activity of different cells within the retina. In 1908, Einthoven and Jolly[4] reported on three components of the ERG. An initial negative deflection was designated by the letter "a"; a subsequent positive component, normally greater in amplitude, was termed "b"; and a final, prolonged positive component was referred to as "c." In 1933, Granit,[5] working with the dark-adapted, rod-dominated cat retina, demonstrated that the ERG consisted of three processes, appropriately called P-I, P-II, and P-III. The Roman numerals designated the order in which a gradual increase in ether narcosis suppressed the various ERG components in the cat. The P-I, P-II, and P-III processes of Granit correspond to the c-, b-, and a-waves, respectively, by which the components of the ERG are now conventionally named. Subsequently, the P-III component was found to consist of two separate components, or phases, that arise from two different classes of retinal cells. The initial phase (termed the receptor potential or fast P-III), the leading edge of which forms the a-wave, reflects the activity of photoreceptor cells and arises from light-evoked closure of sodium channels along the plasma membrane of receptor cell outer segments. The second, more

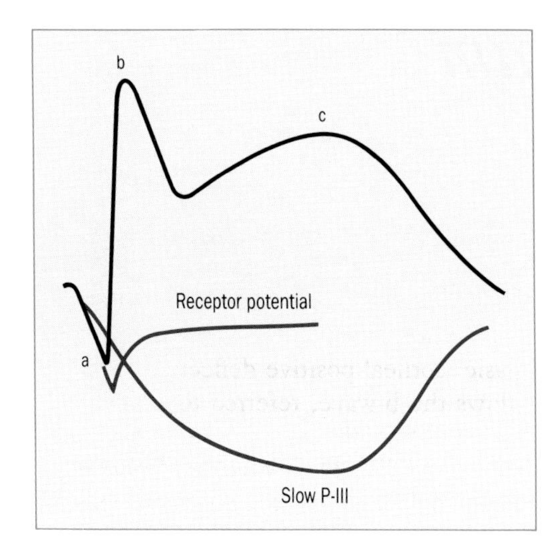

Figure 1-1 *Illustrated are the a-, b-, and c-waves of the ERG. Also shown is the receptor potential, which originates from within the photoreceptor cells, and the slow P-III potential, whose origin is likely in Müller cells.*

Modified by permission from Ripps H, Witkovsky P: Neuron-glia interaction in the brain and retina. In: Osborne NN, Chader GJ, eds. Progress in Retinal Research. *Elmsford, NY: Pergamon Press, 1985; 4:181–219. Copyright 1985, Pergamon Press PLC.*

slowly developing phase (termed the slow P-III), has its likely origin in Müller cells. This response of the Müller cells is thought to be stimulated by a light-induced decrease in extracellular potassium in the vicinity of photoreceptor cell inner segments developing as a consequence of photoreceptor cell activity (Figure 1-1).

1-1

COMPONENTS AND ORIGINS OF THE ERG

1-1-1 a-, b-, and c-Waves and Off-Responses

The site of origin of the ERG components has been a subject of investigation for more than a quarter century. Various approaches for investigation have included comparing the results from (1) the predominantly cone or rod retinas of various species, (2) chemical destruction with toxins known from histologic data to destroy specific retinal cells,[6] (3) microelectrode investigations of individual cell layers, and (4) observations of the ERG components' variation with various stimulus intensities and adaptation levels.

In vertebrate retinas, the absorption of light by visual pigment within photoreceptor outer segments, which in rods is termed rhodopsin, initiates a sequence of not yet fully understood molecular events localized to these segments that generates a wave of hyperpolarization of the photoreceptors. Photoactivated rhodopsin activates a protein, called transducin (or G-protein), which in turn activates the enzyme cGMP phosphodiesterase (PDE), ultimately resulting in a reduction in photoreceptor cell cyclic GMP levels. This reduction in cyclic GMP levels causes the closure of channels in the outer segment membrane, which in darkness is permeable to sodium ions. A hyperpolarization (negative change in intracellular electrical potential) results from the light-induced decrease of the inward-directed sodium current across the plasma membrane of photoreceptor outer segments (more precisely, a decrease in sodium conductance of the plasma membrane). The electrical change thus generated can be measured as the corneal negative a-wave of the ERG. The light-induced hyperpolarization of photoreceptor cells diminishes the release of a neurotransmitter at their synaptic terminals. This modulation of neurotransmitter release in turn causes a depolarization or hyperpolarization of the postsynaptic bipolar and horizontal cells. As a consequence of mainly bipolar cell depolarization, an increase in extracellular potassium, primarily within the postreceptoral outer plexiform layer, causes a depolarization of Müller (glial) cells.[7] The resulting transretinal current flow along the length of the radially oriented Müller cells appears to be largely responsible for the corneal positive b-wave of the clinical ERG.[8-10] A more proximal increase in extracellular potassium, at the level of the inner plexiform layer, following light stimulation appears to derive from depolarization of amacrine cells, bipolar cells, and ganglion cells.[11] The more distal increase in extracellular potassium in the outer plexiform layer contributes substantially more to the extracellular current, and thus to the b-wave, than does the proximal potassium increase.[12]

A decrease in extracellular potassium around photoreceptor cell outer segments following light absorption also alters a standing electrical potential that exists between the apical and basal surfaces of retinal pigment epithelial (RPE) cells. A flash of light induces a transient hyperpolarization at the apical surface of RPE cells and a hyperpolarization of Müller cells, which combine to form a monophasic, corneal positive deflection that follows the b-wave, referred to as the c-wave. Thus, the c-wave represents the algebraic summation of a corneal positive component, resulting from a change in the transepithelial potential induced by hyperpolarization at the apical membrane of the RPE cells, and a corneal negative component generated by hyperpolarization at the distal portion of the Müller cells, the so-called slow P-III.[13] Diffuse degeneration or a malfunction of RPE cells reduces the c-wave amplitude of the ERG. Similarly, diffuse photoreceptor cell degeneration causes a reduction in ERG c-wave amplitude. Ample evidence exists that the c-wave, although reflecting alteration of electrical activity at the level of RPE cells, depends on rod receptor cell activity. The c-wave has the rod spectral sensitivity, is lost when the retina is light-adapted, and is absent in cone-dominant retinas.[14] However, Tomita et al[15] have reported a cone-specific c-wave found in cone-dominant retinas.

The c-wave, which in humans is recorded with a fully dilated pupil in a dark-adapted eye, generally peaks within

2–10 seconds following the onset of a flash stimulus, depending on flash intensity and duration. Both the c-wave amplitude and the time from stimulus to response peak increase with stimulus intensity or stimulus duration. Since the response develops over several seconds, it is subject to interference from electrode drift, eye movements, and blinks. Special recording electrodes, DC amplification (unlike AC amplification, which is used for standard ERG recordings), and both high-luminance and extended-duration light stimuli are necessary for an optimal recording of the ERG c-wave.[16] These procedural constraints, in addition to the wide physiologic variability in c-wave amplitude (some normals appear to entirely lack a recordable c-wave) and waveform, have limited the clinical application of c-wave measurements.

Not readily recognized is the observation that an off-response, resulting from the abrupt cessation of retinal illumination, contributes appreciably to the photopic b-wave amplitude. The traditionally very brief strobe-flash stimuli used for ERG recordings preclude isolation of that portion of the off-response that is part of the photopic b-wave amplitude. However, with more prolonged flashes, it can be made apparent that a substantial portion of the photopic b-wave is actually an off-response component, d-wave (Figure 1-2). Hyperpolarizing bipolar cells,[11] probably in addition to other cells such as the photoreceptors,[17] are likely to contribute to the off-response component of the b-wave amplitude, while depolarizing bipolar cells likely contribute to the "true" (on-response) photopic b-wave by increasing the concentration of extracellular potas-

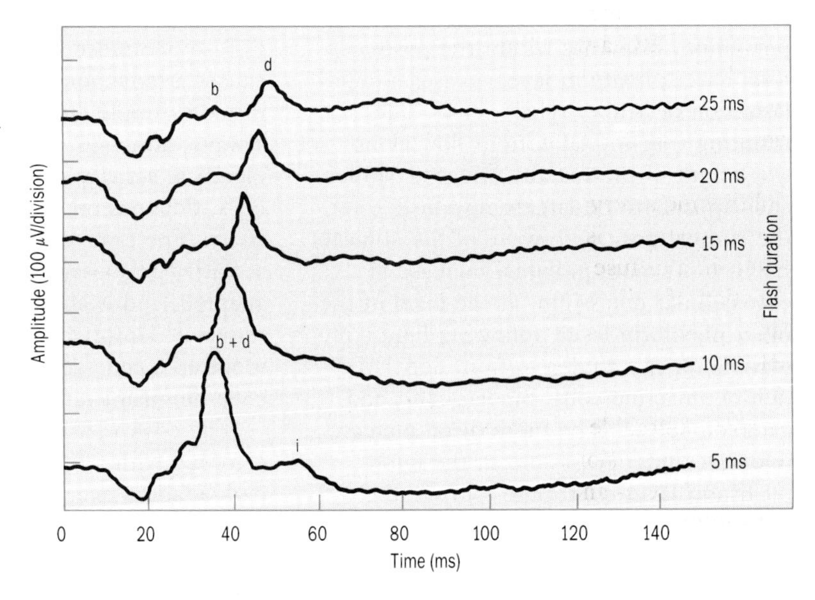

Figure 1-2 *With longer-duration flashes, the photopic b-wave can be demonstrated to contain a substantial off-response (d-wave). The i-wave is also an off-response component.* *Courtesy of Neal S. Peachey, PhD.*

sium. A small positive wave at the end of the descending portion of the ERG b-wave response to a high-luminance flash is also an off-response component, which has been designated as the "i-wave" to suggest the character of the wave as the result of interference between on- and off-response components (Figure 1-2).[18]

Electrical events within ganglion cells or optic nerve fibers do not contribute to the flash-elicited a- or b-waves of the ERG. Thus, disorders such as glaucoma and various types of optic atrophy, which selectively affect ganglion cells and/or optic nerve axons, do not ordinarily reduce ERG a- or b-wave amplitudes. Although isolated reports on both animals and humans show that ERG amplitudes increase subsequent to section of the intracranial portion of the optic nerve,[19,20] this has not been a consistent finding.[21] Since the ERG b-wave necessarily depends on electrochemical events that generate the ERG a-wave, any retinal disorder that prevents generation of a normal a-wave will also affect the development of a normal b-wave. Examples include retinitis pigmentosa, retinal detachment, and ophthalmic artery occlusion. The converse, however, is not true. Disorders that result in a diffuse degeneration or dysfunction of cells within the inner nuclear layer (Müller cells or bipolar cells) can selectively decrease the ERG b-wave amplitude without diminishing the ERG a-wave. A notable example is central retinal artery occlusion.

To reduce a- and/or b-wave amplitudes, a disorder must affect a large area of retinal tissue. Thus, focal lesions of the fovea (defined as a region the approximate size of the optic disc, 1.5 mm in diameter, centered at the foveola) do not affect the a- and b-wave amplitudes elicited by a full-field flash stimulus. The work of Armington et al[22] suggests that a loss of one half the photoreceptors across the entire retina may result in approximately a 50% reduction in ERG amplitude. François and de Rouck[23] showed that a macular lesion, up to a size of 3 disc diameters, produced by photocoagulation showed no modification of ERG amplitudes. This finding was in accord with the investigations by Schuurmans et al,[24] who demonstrated in rabbits that photocoagulation of a 20° retinal area in the posterior pole showed no decrease in ERG amplitude. Between 20° and 60° of photocoagulation caused decreases in ERG amplitude that were proportionally related to the destroyed area.

Thus, normal full-field ERG recordings may be obtained in patients with marked impairment in central vision resulting from disorders that affect the visual system at or proximal to the retinal ganglion cells or from photoreceptor cell degeneration limited to the fovea. Conversely, markedly reduced or even nondetectable ERG amplitudes can be found in the presence of 20/20 acuity, as seen in some cases of retinitis pigmentosa.

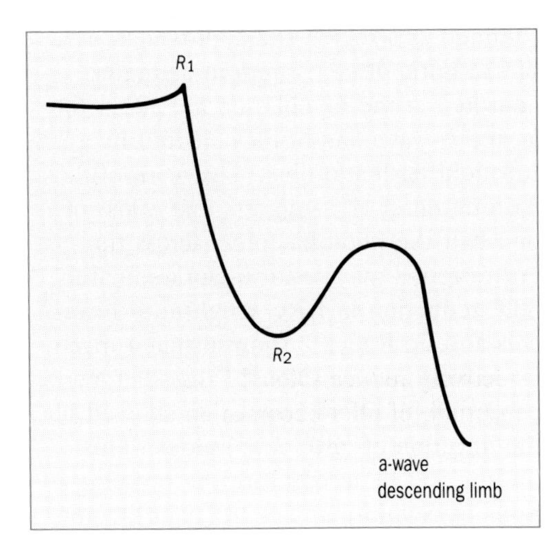

Figure 1-3 *Note the positive (+) R_1 and negative (−) R_2 portions of the early receptor potential (ERP), as described by Brown and Murakami.[25] The ERP occurs immediately after a high-luminance light stimulus. Its negative portion blends with the a-wave of the ERG.*

1-1-2 Early Receptor Potential

The early receptor potential (ERP) is a rapid, transient waveform recorded almost immediately after a light flash stimulus, particularly after one of high intensity in a dark-adapted eye. The response originates from the bleaching of photopigments at the level of the photoreceptor outer segments. This potential, first described by Brown and Murakami[25] in 1964, was found by Pak and Cone[26] to consist of two components, designated R_1 and R_2. The initial, corneal positive portion (R_1) has a peak time of approximately 100 microseconds and appears to be primarily a cone response. The second, corneal negative component (R_2) has a peak time of approximately 900 microseconds and consists of contributions from both rods and cones. This corneal negative component is essentially complete by the time the a-wave begins (Figure 1-3). Both positive and negative components of the ERP resist anoxia. The initial, positive phase can still be observed in the frozen eye, while the second, negative phase disappears on cooling. This potential appears to reflect molecular events within photoreceptor visual pigments and corresponds with photochemical kinetics within these pigments. The corneal positive component, R_1, has been associated with the conversion of lumirhodopsin to metarhodopsin I during the bleaching of visual pigment, while the corneal negative component, R_2, is generated by the conversion of metarhodopsin I to metarhodopsin II.

The ERP is thought to originate as the result of charge displacements that occur within photoreceptor outer segments during the photochemical reactions described above. Its recovery rate has been correlated with the regeneration rates of the visual pigments within the retinal photoreceptors. The human ERP is probably generated predominantly by the cones. By evaluation of cone-deficient subjects, the rod contribution to the ERP has been estimated to be between 20% and 49%.[27,28] The measurement of this response in humans is technically considerably more difficult to record than the ERG and is therefore not seen in routine ERG recordings. It requires the use of a high-intensity flash stimulus, the delivery of the stimulus to the eye preferably in Maxwellian view (requiring the use of a condensing lens and careful optical focusing), and isolation of the recording electrode from the flash stimulus to avoid a photovoltaic artifact that would obviate recording of the ERP response.

1-1-3 Oscillatory Potentials

In 1954, Cobb and Morton[29] first reported the presence of a series of oscillatory wavelets superimposed on the ascending limb of the ERG b-wave after stimulation by an intense light flash. Yonemura[30] subsequently termed these wavelets oscillatory potentials (OPs). These potentials are high-frequency, low-amplitude components of the electroretinogram, with a frequency of about 100–160 Hz, to which both rod and cone systems can contribute.[31,32] By comparison, the a- and b-waves are dominated by frequency components of about 25 Hz.[31] The cellular origin of OPs within the retina is uncertain, although it is likely they are generated by cellular elements other than those that generate the a- and b-waves. Current information supports the notion that cells of the inner retina, supplied by the retinal circulation, such as the amacrine or possibly interplexiform cells, are the generators of these potentials. Intravitreal injection of glycine, which produces morphologic changes in amacrine cells,[33] results in a loss of OPs. Other neurotransmitters, including gamma-aminobutyric acid (GABA), glutamate, and dopamine, can selectively reduce OPs.[34,35] Reduction in amplitude of OPs becomes apparent in the presence of retinal ischemia, such as that seen in patients with diabetic retinopathy, central retinal vein occlusion, and sickle cell retinopathy. Reduction in OP amplitudes has also been reported in patients with X-linked juvenile retinoschisis,[36] in some patients with congenital stationary night blindness,[37] and in patients with Behcet's disease.[38]

Technically, OP responses are recorded with procedures similar to those used for standard a- and b-wave amplitudes. However, for optimal isolation, the high-pass filter setting on the amplifier should be set at about 100 Hz rather than about 1 Hz, which is frequently used for recording the lower-frequency a- and b-waves. Further, a "conditioning" flash should be presented approximately 15–30 seconds prior to averaging a subsequent set of 3 or 4 responses to additional flashes at about 15-second intervals from a dark-adapted eye. Responses obtained under these stimulus conditions are from the cone system (the conditioning flash having adapted the rod system so that it does not contribute to the subsequent flash-elicited OPs).[32] Os-

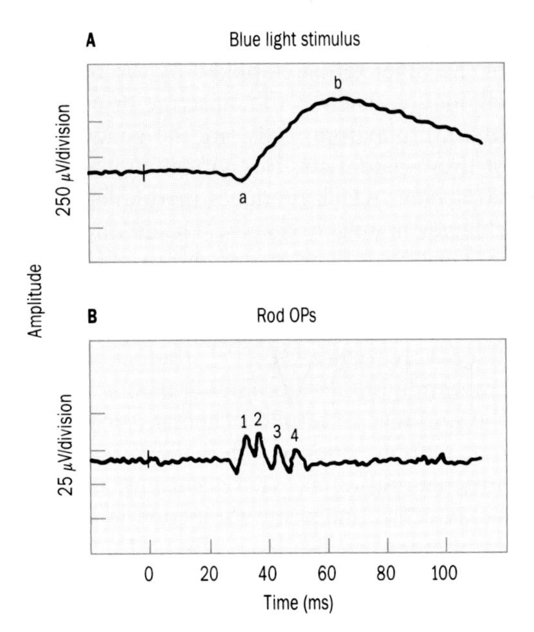

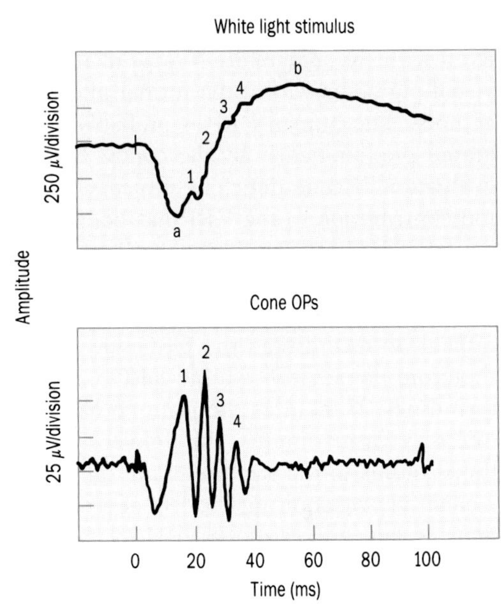

Figure 1-4 *(A) Single-flash and (B) oscillatory-potential responses to a blue and a white light stimulus in a dark-adapted eye. OPs to the blue stimulus are from the rod system, while those to the white stimulus are from the cone system.* *Courtesy of Neal S. Peachey, PhD.*

cillatory potentials from the rod system can be obtained with low-intensity blue light (Figure 1-4). Although OPs are generated by cells of the inner retina, they will be reduced in amplitude by disorders that affect more distal retinal cells, since the distal cells provide electrical signals that form the input to the more proximal generating cells. Thus, diseases that seem to affect primarily the outer retina, such as cone dystrophy or retinitis pigmentosa, reduce OPs as well as a- and b-wave amplitudes.

1-2

MEASUREMENT OF THE ERG COMPONENTS

An evaluation of ERG components includes the measurement of both amplitudes and timing characteristics. Latency

A

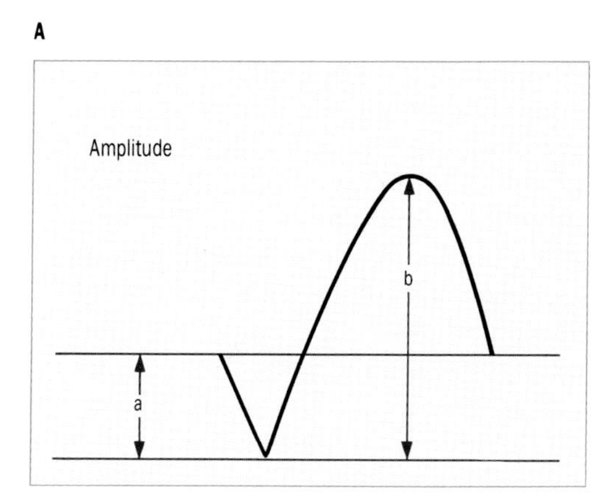

B

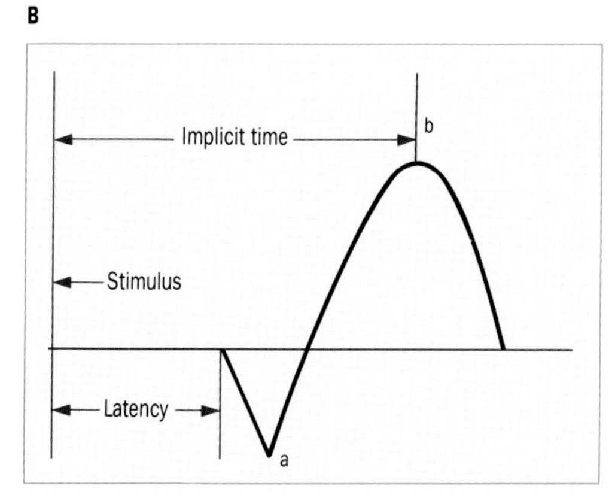

Figure 1-5 *Methods of measuring ERG (A) waveform amplitudes and (B) time relationships. Most examiners measure b-wave amplitude from the trough of the a-wave to the peak of the b-wave. Latency of response refers to the time from the stimulus until the beginning of the ERG a-wave, while implicit time (b-wave response illustrated) is measured from the stimulus onset to the peak of the b-wave.*

(the time from the stimulus until the beginning of the response), implicit time (the time from the onset of the light stimulus until the peak amplitude response), and amplitude are all depicted in Figure 1-5. Note that the a-wave peak amplitude is measured from the baseline to the trough of the a-wave, while the b-wave is traditionally measured from the trough of the a-wave to the peak of the b-wave. If only the a- and b-waves are included, the duration of the ERG response is less than 0.25 second. Table 1-1 lists the normal ranges of a- and b-wave amplitudes and implicit times to single-flash stimuli under scotopic (dark-adapted) and photopic (light-adapted) conditions. These values vary with the intensity of the light stimulus, the state of retinal adaptation, and several other variables described in Sections 1-5 and 1-6.

The ERG responses of the right and left eyes of normal subjects under controlled standard conditions using a *ganzfeld* (full-field) stimulus are, on the average, usually within 10% of each other, with a maximal difference of approximately 20–

TABLE 1-1

ERG Values With Maximum-Intensity Stimuli (Approximate Values)

	Photopic	Scotopic
a-wave amplitude	20–50 μV	190–300 μV
a-wave implicit time	14–20 ms	20–26 ms
b-wave amplitude	90–180 μV	400–700 μV
b-wave implicit time	26–34 ms	40–56 ms

24%. The 24% represents the maximal difference caused by shifts in the position of a contact lens even under controlled conditions. Thus, a difference in amplitude between the two eyes of a patient, or the same eye on different occasions, is probably pathologic if it is between 20% and 24% and highly likely to be pathologic if it exceeds 24%. Some investigators indicate that a reduction in ERG amplitude during serial follow-up examinations is significant, at the 99% confidence limit, if a decline of greater than 31% occurs to a single-flash white stimulus and greater than 44% to a 30-Hz flicker stimulus.[39]

1-3

RECORDING PROCEDURE

The most basic aspect of obtaining an ERG is an adequate light stimulus that allows a variability in stimulus intensity over a range suitable for clinical electroretinography. The specific model or characteristics of the stimulus are probably less important for clinical use than the ability to calibrate the stimulus and maintain it at a constant luminance. Many examiners use a Grass xenon-arc photostimulator, with a flash duration of 10 microseconds.

A contact-lens electrode with a lid speculum is most often used to record retinal responses to the light stimulus. A topical anesthetic is administered to reduce any difficulty and discomfort of lens insertion. A ground electrode is generally placed on the patient's earlobe with electrode paste. A reference, or "inactive," electrode is placed centrally on the patient's forehead slightly above the supraorbital rims or, alternatively, on the mastoid region or earlobe. The reference electrode serves as the more negative pole (having been placed closer to the electrically negative posterior pole of the eye). In the Burian-Allen bipolar electrode, a silver reference electrode is incorporated within the lid speculum, obviating the need for a forehead reference electrode. The corneal contact lens (active electrode), ground electrode, and reference electrode all connect with a junctional box, from which the signals are delivered to additional recording components for amplification and, finally, display. Most often, display and amplification systems are contained within a single unit. A differential amplifier is used to record the difference in voltage between the active corneal electrode and the inactive reference electrode that occurs following light stimulation of the retina.

1-4

RECORDING ELECTRODES

Important features of ERG recording electrodes include (1) quality in their components that facilitates stability of responses with low intrinsic noise levels, (2) patient tolerance with limited irritation of the corneal surface, and (3) availability at a reasonable cost.

1-4-1 Burian-Allen Electrode

The Burian-Allen contact-lens electrode has already been mentioned briefly.[40] Lens sizes suitable for adults to premature infants are available. These lenses have the advantage of a lid speculum, which promotes a more consistent input of light entering the pupil by controlling, or at least limiting, the effects of blinking and lid closure during photic stimulation. Images viewed through the central portion of the lens, made of polymethylmethacrylate, show variable degrees of blur. Thus, the lens is not optimal for use when stimulus image clarity is vital; eg, the pattern ERG. The recording electrode consists of an annular ring of stainless steel that surrounds the central polymethylmethacrylate contact-lens core. In the bipolar version of the Burian-Allen lens, a conductive coating of silver granules suspended in polymerized plastic is painted on the outer surface of the lid speculum. This conductive coating serves as a reference electrode, obviating the need for an additional separate Ag-AgCl wire, as is necessary with the unipolar model.[41]

1-4-2 Dawson-Trick-Litzkow Electrode

The Dawson-Trick-Litzkow (DTL) electrode consists of a low-mass conductive Mylar thread whose individual fibers (approximately 50 microns in diameter) are impregnated with metallic silver. The conductive thread virtually floats on the corneal film surface.[42]

When compared to the Burian-Allen electrode, the amplitudes of the ERG responses are on average 10–13% less with the DTL electrode. However, the variance in signal amplitude with the DTL electrode was found to be less than with the Burian-Allen electrode. In addition, its noise characteristics were favorable, as was its tolerance for extended recording periods.[43]

Major advantages of the DTL electrode include its low cost, patient acceptance, and clear imagery. However, blinks after a flash can cause a movement artifact, which sometimes will obscure ERG components.

1-4-3 ERG-Jet Electrode

The ERG-jet electrode lens is made of a plastic material that is gold-plated at the peripheral circumference of its concave surface.[44] Its advantages include sterility (the lens is packaged and can be discarded after each use), simplicity in design, and ease of insertion, which has potential benefit for sensitive eyes. Movement of the lens during recordings and absence of a lid speculum could contribute to variability of recordings in individual instances.

1-4-4 Mylar Electrodes

Aluminized or gold-coated Mylar has also been used for recording ERG potentials.[45,46] The gold-foil electrode is preferable because it provides a more stable baseline and is less likely to dislodge or flake than the aluminum Mylar.[47] The gold-foil electrode, which fits over the lower eyelid, is less sensitive to eye movement than is the DTL electrode, yet more sensitive to such movement than are contact-lens–style recording lenses. Both of these electrodes are probably unsuitable for DC amplification measurements because of rather large, low-frequency drift. Borda et al[47] reported a lower amplitude, by 34–44%, with a gold-coated Mylar electrode than that recorded with a contact-lens electrode.

1-4-5 Skin Electrodes

For specific purposes—for instance, in recording ERGs from infants and small children—the corneal electrode can be replaced by an electrode placed on the skin at the lower eyelid. Recorded amplitudes are 10–100 times smaller than those from a corneal ERG.[48,49] Both the lower amplitudes and the variability in responses with the use of skin electrodes preclude their use for purposes other than screening. Whenever possible, it is preferable to use a smaller recording electrode with a lid speculum, such as an infant Burian-Allen electrode, which can be used successfully in children and infants.

1-4-6 Cotton-Wick Electrodes

Sieving et al[50] introduced an electrode consisting of a Burian-Allen electrode shell fitted with a cotton wick. This electrode, as well as earlier variants of a cotton-wick configuration, is of value in recording ERG amplitudes to a high-luminance flash and for recording the ERP (early receptor potential), since the lens is free of a photovoltaic artifact, which can occur when intense light illuminates the metal present in other recording electrodes.

1-5

THE ERG UNDER PHOTOPIC (LIGHT-ADAPTED) AND SCOTOPIC (DARK-ADAPTED) CONDITIONS

The recording of the photopic (cone) ERG potential can be done before or after dark adaptation, with the patient light-adapted to a background light intensity of approximately 7–10 foot-lamberts, which facilitates the recording of a response exclusively from the cone system. Regardless of which sequence is adopted, the response obtained to a high-luminance stimulus (Figure 1-6) is relatively small in amplitude, with a short implicit time and a brief duration. The amplitude of the cone b-wave grows with stimulus intensity and tends to approach a maximal limiting value at the highest intensities. The cone b-wave implicit time increases slightly with increasing stimulus luminance.[51]
If single-flash or 30-Hz flicker cone responses are obtained after a period of dark adaptation, it is vital to wait approximately 10–12 minutes for light adaptation before obtaining final photopic amplitudes, since there is a progressive increase in ERG amplitude, the extent of which depends on the level of the background adapting light and the stimulus intensity.[51-57] More intense backgrounds and stimuli will be associated with greater increases in cone ERG amplitudes during the period of light adaptation. Under certain stimulus and background intensities, amplitudes can approximately double (Figure 1-7). A progressive decrease in cone b-wave implicit time may also be noted during the course of light adaptation.[56] In normals, a rod–cone interaction appears to influence the light-adapted cone b-wave implicit time. Light adapta-

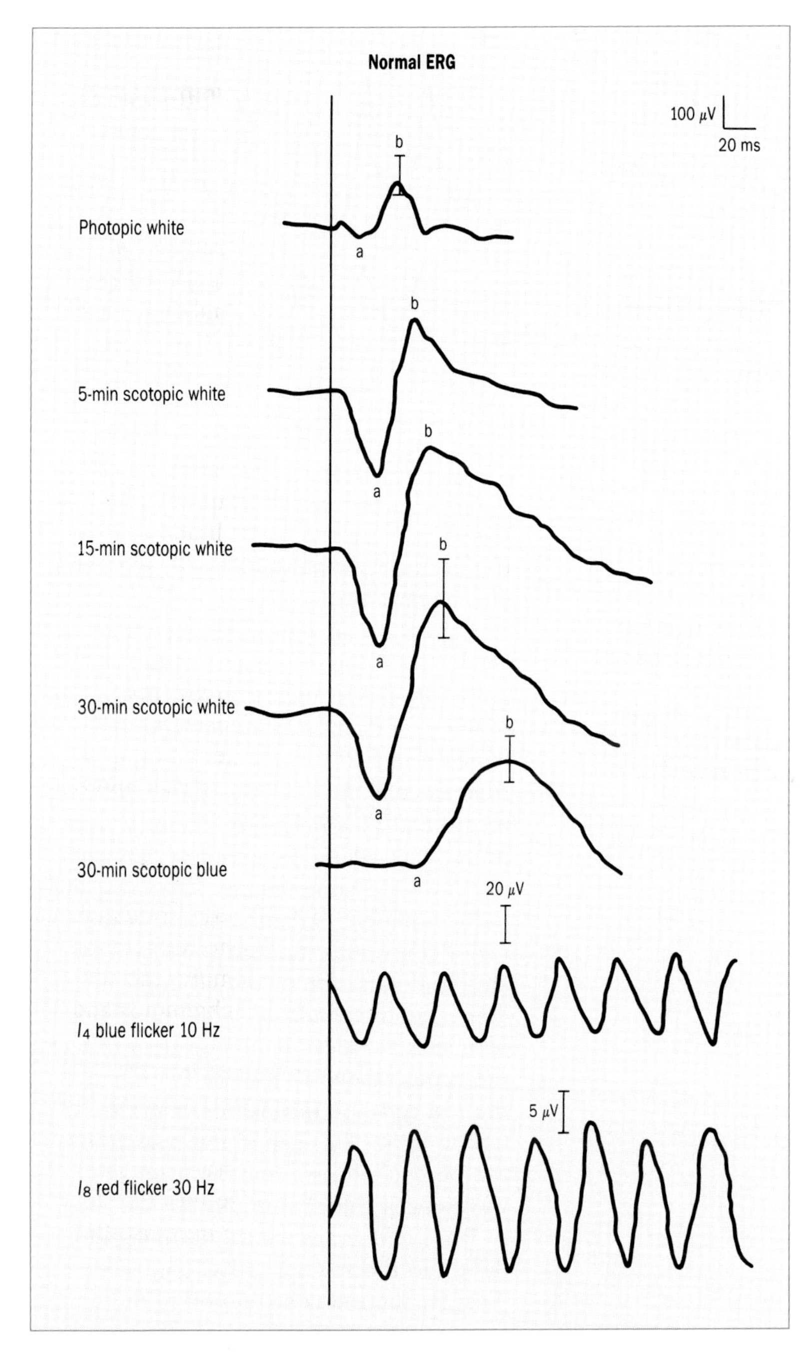

Figure 1-6 *Normal ERG responses under photopic and scotopic conditions. Cone responses are generally isolated under proper photopic conditions and with use of a 30-Hz flickering stimulus, while rod function can be isolated with use of a low-luminance, short-wavelength (blue) stimulus as either a single flash or flickering at 10 Hz. A high-luminance white light flash under dark-adapted conditions measures a combined rod and cone response that in the normal subject is predominantly from the rod system. The terms I_4 and I_8 refer to the stimulus intensity. In this as well as subsequent figures, the time calibration of 20 ms refers only to single-flash recordings.*

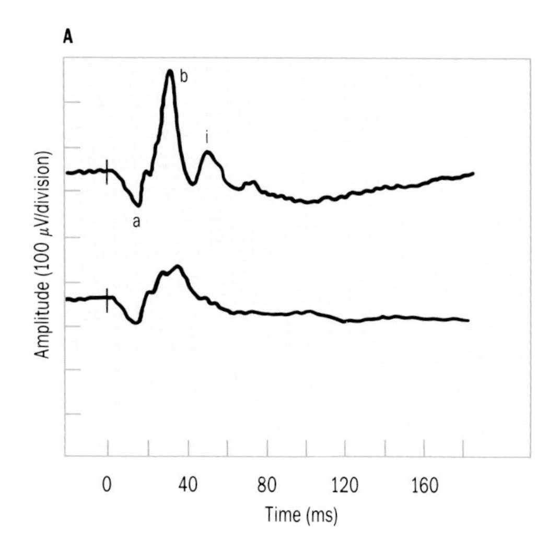

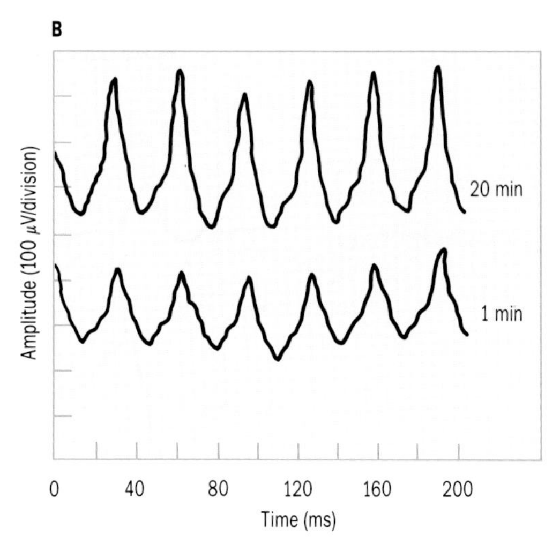

Figure 1-7 *(A) Light-adapted single-flash white and (B) 30-Hz white flicker cone responses at 1 minute and 20 minutes of light adaptation subsequent to a period of approximately 30 minutes of dark adaptation. Note the importance of allowing for a suitable period of light adaptation to properly determine maximal cone amplitude responses. Also note changes in implicit time during light adaptation.*

Courtesy of Neal S. Peachey, PhD.

tion of the rods shortens the implicit time of the cone b-wave.[58]

Figure 1-6 also shows scotopic (dark-adapted) responses elicited with a white light stimulus of high luminance after 30 minutes of dark adaptation. Note the increase in ERG amplitude and implicit time as compared to the light-adapted recording. This response, although evoked under dark-adapted conditions, has both rod and cone components. However, the rod component is dominant and is the major contributor to both the increased amplitude and the increased implicit time. This is understandable since the rod population is appreciably greater than that of the cones (about 17 rods to 1 cone) and the rod cells are intrinsically more sensitive to light. An isolated rod response can be evoked by a low-intensity short-wavelength (blue) stimulus (Figure 1-6).

Rods and cones contribute independently to the scotopic ERG amplitude. That is, the rod contribution to the scotopic b-wave is the same whether the cone contribution is present or absent. As will be emphasized in subsequent sections, excluding the influence of retinal or choroidal disease, those recording conditions that determine the relative cone and rod contributions to the ERG response include the intensity, wavelength, and frequency of the light stimulus, in addition to the state of retinal adaptation. Therefore, a comprehensive evaluation of retinal function includes obtaining ERG responses under testing conditions that include (1) evaluation of the ERG with a constant stimulus intensity during dark adaptation, (2) evaluation of the dark-adapted retina with various stimulus intensities, (3) the ERG response to different-wavelength stimuli, (4) the ERG response to different stimulus frequencies, and (5) the ERG response with the retina light-adapted.

1-5-1 During Dark Adaptation With a Constant-Intensity Stimulus

Figure 1-8 is a schematic representation of the ERG during dark adaptation with the use of a high-luminance flash. Traditionally, the b-wave includes two subcomponents, b_1 and b_2. Note the immediate increase in amplitude of b_1 from that seen under photopic conditions; it occurs even before 1 minute of dark adaptation. The initial growth of b_1 amplitude is likely related to a "neural" network adaptation of primarily the rod response from the adapting light used under photopic conditions and/or a receptoral adaptation within the outer segments. Subsequently, both b_1 (a combined cone and rod response when elicited by a moderate or high-intensity flash stimulus) and b_2 (a presumed rod

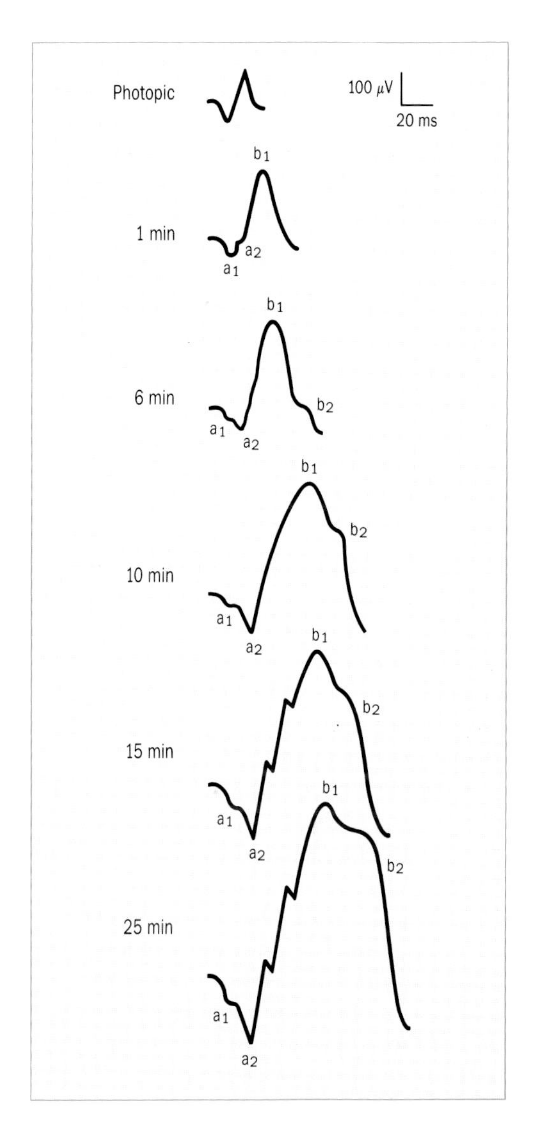

Figure 1-8 *ERG responses during dark adaptation with use of a high-luminance stimulus. Increases in both a- and b-wave amplitudes with progressive dark adaptation are apparent. The b_1 wave has both rod and cone system components, while the b_2 wave is likely exclusively a rod system component. Oscillatory potentials are also more apparent with progressive dark adaptation.*

component of the b-wave) increase in amplitude as the photoreceptor cells dark-adapt and become progressively more sensitive. The rate at which b_2 develops relates to the duration and intensity of previous light exposure. The major growth in b_1, as well as the entire increase in b_2 amplitude, occurs because of the increase in rod receptor response. Note that the implicit time of both a- and b-waves also increases as the ERG changes from a light-adapted to a dark-adapted response. Also note the presence of a biphasic a-wave with the high-luminance flash as well as the appearance of oscillatory potentials on the ascending limb of the b-wave during progressive dark adaptation. It is likely that an oscillatory potential divides the a-wave into its a_1 and a_2 components.[59] The a-wave amplitude also progressively increases during dark adaptation as photoreceptor cells gradually recover their sensitivity.

Examiners need to be aware of an artifact that may be present, particularly with the use of a bipolar recording electrode, in the clinical ERG of some patients. This artifact occurs with a latency that is fast enough to interfere with the scotopic b-wave of the ERG. Since most of the artifact is probably due to a reflex contraction of the orbicularis muscle, it has been termed the photomyoclonic reflex.[60] This artifact is less likely to be apparent with the use of unipolar than with bipolar lenses.

1-5-2 Dark-Adapted Retina With Variable-Intensity Stimuli

Figure 1-9 shows dark-adapted ERG responses to flash stimuli of minimal, moderate, and high luminance. Generally, the amplitudes of both a- and b-waves increase and the implicit times decrease as a function of stimulus luminance; in both cases the curves approach a plateau at high stimulus intensities. It is noteworthy that the b-wave implicit time appears to *increase* with stimulus intensity under light-adapted conditions. Figure 1-10

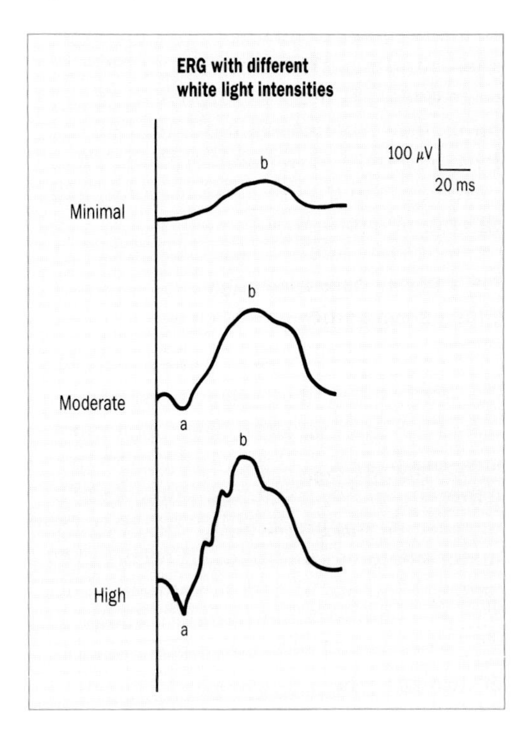

Figure 1-9 *ERG responses in a dark-adapted eye with stimuli of different white light intensities. Note the shorter implicit times and larger amplitudes as stimulus intensity is increased.*

shows a series of ERG responses to flashes of increasing intensity. The lowest-intensity flashes elicit an exclusively rod response, while high- and medium-intensity stimuli evoke responses that, as already noted, are rod-dominant but contain both rod and cone components. Note the presence of oscillatory potentials at the higher stimuli and the absence of the a-wave at the lower stimulus intensities. Because there is a large increase in gain for the transmission of information from photoreceptor cells to inner retinal neurons, the sensitivity of the b-wave is at least 1 log unit greater than that of the a-wave. Thus, b-wave amplitudes can be measured at low stimulus intensities, where a-wave responses are not yet apparent.

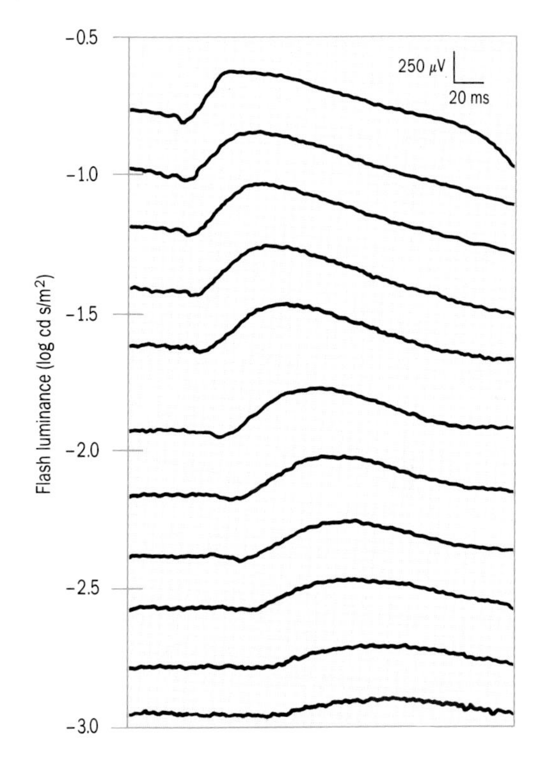

Figure 1-10 *A series of ERG responses to white light stimuli of increasing luminance. The lowest luminance flashes elicit exclusively a rod response, while higher and medium luminance stimuli evoke responses that represent combined rod and cone activity. Note the presence of oscillatory potentials at the higher stimuli as well as the absence of an a-wave at the lowest stimulus intensities.*

Courtesy of Neal S. Peachey, PhD.

Electroretinographic investigations of patients with retinal disorders now more frequently include studies of scotopic b-wave responses to a range of full-field stimulus intensities to obtain a stimulus/response (S/R) function. The normal S/R function (Figure 1-11) exhibits an increasing amplitude over a stimulus range of about 2 log units. The S/R relationship can be summarized by the equation

$$R/R_{\max} = I^n/(I^n + \sigma^n)$$

where R is the b-wave amplitude produced by a flash of intensity I (typically in foot-lamberts per second or cd s/m^2) and σ (sigma) is the value of the flash intensity that produces one half the maximal response $R_{\max}$. The sigma value is an index of retinal sensitivity, whereas the exponent n, which has a value of about 1 for the normal retina, refers to the slope of the S/R function. Stimulus/response

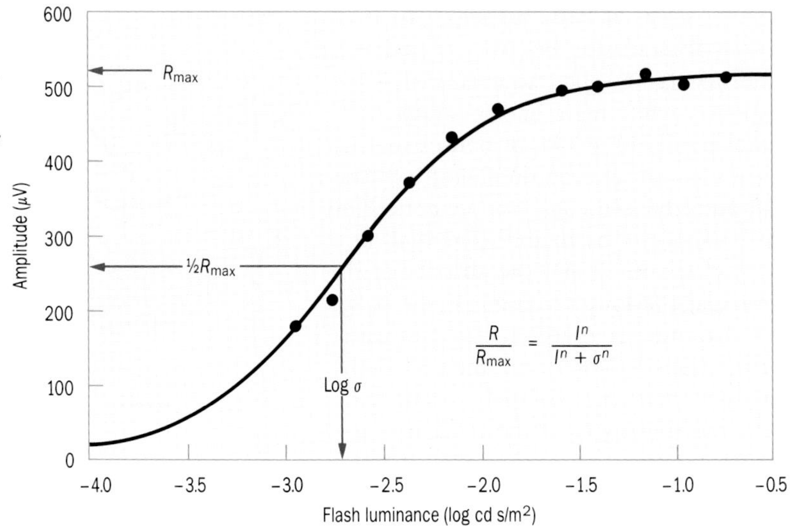

Figure 1-11 *Flash stimulus versus amplitude response function curve with equation used to determine curve fit. See text for description.*

determinations can provide information that has implications for the nature of various retinal degenerative disorders. For example, a total loss of retinal receptor cells in isolated retinal regions would probably reduce R_{max}, but not appreciably affect the value of σ or n. In contrast, Massof et al[61] propose that a notable reduction in the overall concentration of rhodopsin, as might occur if the outer segments of the photoreceptor cells were shortened as a result of a disease process, would not appreciably affect R_{max}, but would decrease the retinal sensitivity; ie, the value of σ.

1-5-3 With Different-Wavelength (Color) Stimuli

As already noted, a low-intensity blue stimulus can elicit an ERG response that is exclusively from the rod system. Figures 1-6 and 1-12 illustrate this response in a fully dark-adapted eye. The low-intensity stimulus precludes stimulation of significant cone activity, while the use of a broad-band filter that transmits light in the short-wavelength (blue) region of the spectrum (less than 460 nm) maximizes stimulation of the rods. A long-wavelength (orange-red) light (about 600 nm) elicits a characteristic biphasic positive response (Figure 1-12). The early portion of the response, referred to in the literature as the x-wave,[62] is ascribed to cone activity. This peak is missing in protanopes and protanomals[63] (red defectives) and achromats (those totally color-blind, also called monochromats). The late peak ascribed to rod activity is absent or markedly reduced in, among others, patients with congenital stationary night blindness. Blue and orange-red or red stimuli can be "scotopically balanced" to produce equal amplitudes from the dark-adapted eye. By digital subtraction of the matched responses, the rod contribution can be

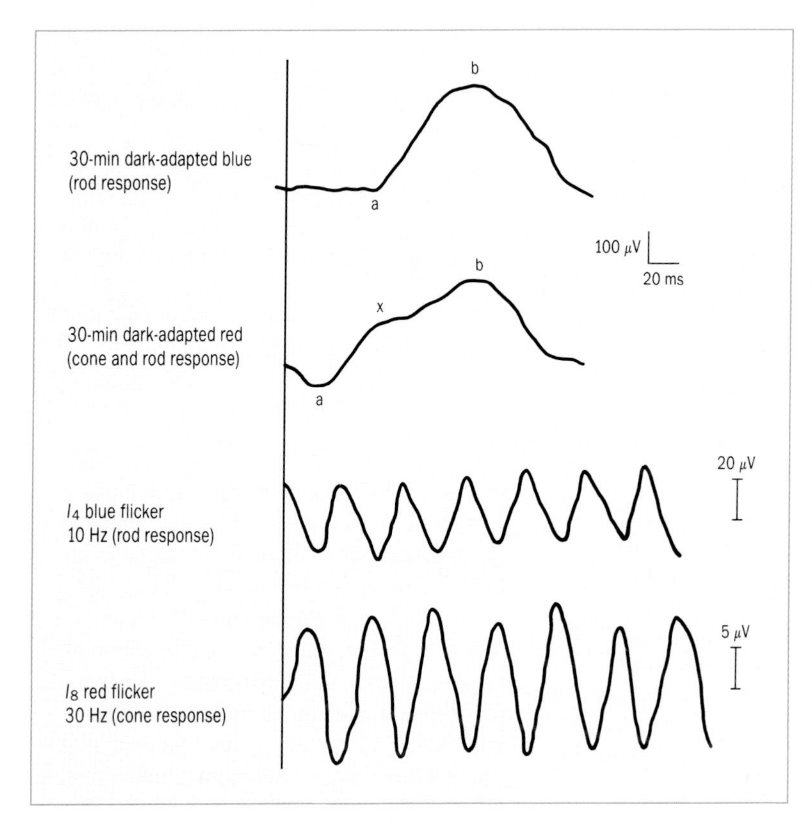

Figure 1-12 *ERG responses to either single-flash or flicker chromatic stimuli. Low-luminance blue stimuli measure isolated rod function, while 30-Hz red or orange-red flicker stimuli determine cone function. Both cone (x-wave) and rod (b-wave) function can be determined with use of a single-flash long-wavelength (red) stimulus after dark adaptation.*

eliminated, thereby facilitating the determination of dark-adapted cone function.[51] In addition to a single-flash stimulus, intermittent chromatic stimuli can be used to elicit 10-Hz blue or 30-Hz red flicker responses to obtain isolated responses from the rod or cone systems, respectively (Figure 1-12).

1-5-4 With Variable-Frequency Stimuli

As the frequency of retinal stimulation with light of moderate intensity increases from 1 to 70 flashes per second, the various types of retinal receptors become incapable of responding at the higher frequencies. The stimulus frequency at which separate responses can no longer be recorded is the flicker fusion frequency (Figure 1-13). Up to approximately 20 flashes per second, the rods can respond with a separate response for each flash. On the other hand, the cones can generate separate responses to flash stimulus rates as high as 70 flashes per second. The flicker-evoked responses indicated in Figures 1-12 and 1-13 represent electrical activity from both photoreceptor cells, which generate the a-wave, and more proximal retinal cells, which generate the

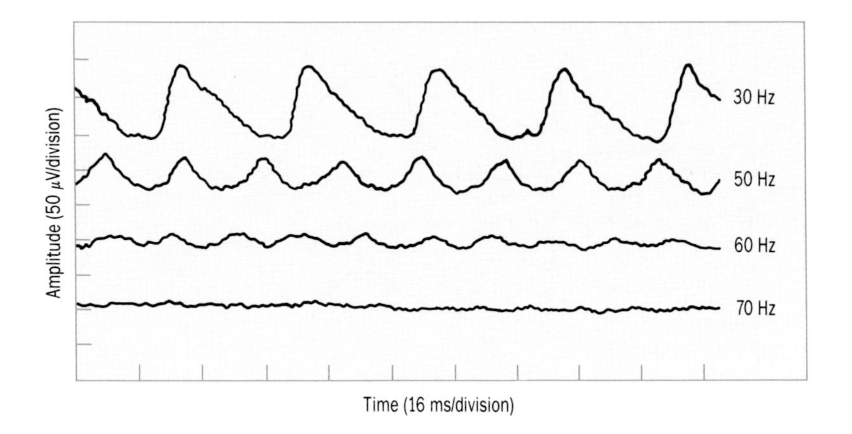

Figure 1-13 *With intensity held constant, light flashes of increasing frequency will record responses that progressively decrease as stimulus frequency is increased. A frequency is reached when individual responses are no longer recordable (flicker fusion frequency). Individual responses can be elicited in normal subjects with frequencies up to 50–70 Hz.*

b-wave. The greater the luminance of the stimulus, the higher will be the flicker fusion frequency. Measurement of both flicker fusion frequency and the amplitudes seen at faster stimulus frequencies is of value in the diagnosis of diseases with predominantly cone pathology. These responses are of more optimal quality, and better tolerated by the patient, if obtained in the presence of a background light equivalent to that used for obtaining single-flash cone responses.

1-6

OTHER FACTORS AFFECTING THE ERG

1-6-1 Duration of Stimulus

For short-duration flashes, a reciprocal relationship exists between stimulus duration and intensity such that if the intensity is held constant, longer stimulus durations will result in larger ERG amplitudes up to a certain duration limit. In the human eye, some investigators found that responses to stimuli longer than 20 milliseconds are all identical in amplitude.[64]

Johnson and Bartlett[65] determined a longer critical duration of 100 milliseconds in the dark-adapted eye. An increase in stimulus duration beyond this level resulted only in an increased duration of the response but not in a greater amplitude. Note that whether stimuli of 20 or 100 milliseconds are used, an appropriate interval between light flashes of these durations must be maintained to preserve dark-adapted conditions and the consequent reproducibility of amplitudes. As indicated in the discussion of off-responses, if longer-duration stimuli are used under photopic conditions, a major contribution to the b-wave can be demonstrated to consist of an off-response (d-wave). Longer-duration stimuli separate the off-response from the b-wave, revealing its appreciable contribution to the b-wave response that is obtained using the more typical short-duration flashes (see Figure 1-2).

1-6-2 Size of Retinal Area Illuminated

The full-field ERG results primarily from a diffuse retinal response. The surface of the retina and ocular media scatters the stimulus to affect large retinal areas. Thus, a light focused on the optic disc can produce a sizable ERG due entirely to internal scatter. With direct flash methods, in which the light stimulus is not reflected from within a diffusing sphere, the stimulus will illuminate various retinal areas in a nonhomogeneous manner. Thus, the total response is the summation of multiple small ERGs of different amplitudes and implicit times. The overall amplitude will be reduced and the implicit time slightly prolonged. A ganzfeld (full-field) stimulus is obtained when light is evenly diffused from within a sphere and consequently illuminates all retinal areas homogeneously. By this method, multiple ERG responses from the retina that are relatively similar in both amplitude and time course are summated.[66-68] These summated regional responses differ somewhat because of, among other reasons, the lack of retinal spatial isotropism and restrictions on homogeneous retinal illumination imposed by the pupillary aperture.[66]

1-6-3 Interval Between Stimuli

The importance of the interval between successive stimuli, particularly those of high intensity or long duration, was mentioned in Section 1-6-1. In the dark-adapted eye, it is important to keep successive stimuli far enough apart to prevent excessive light adaptation, which will tend to decrease both the amplitude and the implicit time of the response; the precise interval will vary, depending on the duration and luminance of the stimulus. In the light-adapted eye, an interflash interval longer than 50 milliseconds usually is sufficient to avoid attenuation of successive responses. With a shorter interval, responses will tend to partially overlap on the record.[69]

1-6-4 Pupil Size

Retinal illumination is proportional to pupil area. ERG recordings are optimally obtained when the pupils are fully dilated. In the dark-adapted eye with use of a high luminance and brief flash stimulus, pupil size is of only moderate concern except in the presence of significant miosis, as in glaucoma patients receiving miotics. In these patients, a reduction in amplitude might be anticipated as less light reaches the retina. However, in the light-adapted eye, particularly with the use of a flicker stimulus, proper pupil dilation is required.

1-6-5 Circulation and Drugs

Henkes,[70] among others, has shown that agents modifying systemic blood pressure and blood flow can affect the ERG amplitude. Vasodilators, including papaverine, acetylcholine chloride, and tolazoline (*Priscoline*), have produced an increase in b-wave amplitude. Hyperventilation can also cause an increase in ERG amplitude.

1-6-6 Retinal Development

With intense stimuli, both photopic and scotopic ERG components can be recorded within 14 hours of birth, although reports vary from different laboratories as to when recordable responses can first be obtained. Horsten and Winkelman[71] reported that an ERG can be recorded from all normal infants within a few hours of birth and even from premature infants well before term. Implicit times, however, are prolonged as compared to adult values. The photopic component, according to these authors, achieves adult values at approximately 2 months of age, while the maximum b-wave amplitude obtainable under dark-adapted conditions gradually reaches adult levels at approximately 1 year, a finding also noted by others.[72,73] At 2–3 months of age, the mean maximum dark-adapted b-wave is about half that of adults.[73,74] Dark-adapted ERG sensitivity, defined as the flash intensity that produces half the maximum ERG b-wave amplitude, becomes equivalent to that of adults at age 5–6 months.[74,75] Flicker fusion findings resemble adult values at 2–3 months of age, while implicit times reach adult values at about the end of 1 year. Different results from various laboratories investigating ERG development in infants might be anticipated, not only because of different experimental conditions, but also as a consequence of wide individual variability in postnatal evolution of the ERG.

1-6-7 Clarity of Ocular Media

Opacification of the lens, particularly that associated with yellowing of the nucleus with age, can reduce the ERG a- and b-wave amplitudes and prolong implicit times as the consequence of its filter effect and a reduction of the amount of light reaching the retinal photoreceptors. Similar reduction in amplitude and prolongation of implicit times can be seen with hemorrhage in the vitreous. Less extensive opacities within the lens cortex can scatter or diffuse light to such an extent that ERG amplitudes and implicit times will not be adversely affected.

1-6-8 Age, Sex, and Refractive Error

Both age and refractive errors are associated with a measurable decrement in the ERG amplitude. Peterson[76] noted a linear reduction with age of the b-wave amplitude, apparently the same for both sexes, from about the age of 10. An exception was noted in females 40–49 years of age, in whom a significant increase occurred, possibly related to hormonal factors. In all age groups, women had a statistically significant higher value of the b-wave amplitude—a finding also noted, at least in some age groups, by Vainio-Mattila[77] and subsequently by Zeidler.[78] Even early ERG investigations by Karpe et al[79] disclosed a reduction in ERG amplitude with age and higher amplitudes in females, particularly in patients above age 50. A similar age-related reduction in ERG amplitude, for at least some components of the response, was noted by Zeidler,[78] Martin and Heckenlively,[80] and Weleber.[81] No age-related changes in photopic or scotopic implicit times were reported by Weleber.[81] In a report by Lehnert and Wunsche,[82] a significant reduction in b-wave amplitude did not occur until the third decade. In high myopes (6 diopters or greater), a reduction in the b-wave amplitude can be observed.[76,83] The responses are usually minimally to moderately subnormal, but may occasionally be markedly subnormal. Apparently, an increase in axial length, independent of any associated chorioretinal degeneration, is an important contributing factor.[83]

1-6-9 Anesthesia

Anesthesia may affect the b-wave amplitude to variable degrees, depending in part on the nature of the drug. Some laboratories report up to a 50% reduction in scotopic b-wave amplitude with certain drugs. The photopic responses and scotopic a-wave amplitudes are generally either minimally affected or not altered. ERG changes also occur in patients during inhalation of nitrous oxide. Under light narcosis, an increase in the b-wave amplitude is noted, while deep narcosis induces a decrease in the b-wave potential. Presumably, with most general anesthetics, the ERG amplitude depends on the level of narcosis. In young children sedated with a barbiturate such as thiopental (*Pentothal*), appreciable reductions in ERG amplitudes have been seen, whereas ketamine seems not to have a notable effect on amplitudes. Padmos and van Norren observed a prolonged cone dark adaptation, measured by ERG, with the use of halothane anesthesia; this effect was attributed to a delay in cone pigment regeneration.[84,85] A similar prolongation in cone dark adaptation was seen with the use of other volatile anesthetics such as methoxyflurane, chloroform, and diethylether.[86] These authors showed that rod as well as cone dark adaptation was prolonged with the use of halothane anesthesia.[87]

1-6-10 Diurnal Fluctuations

A diurnal variation of the rod ERG amplitude has been noted with an approximately 13% reduction in b-wave amplitude occurring $1\frac{1}{2}$ hours after the onset of daylight, which corresponds to the time of maximal rod outer segment disc shedding.[88] Thus, serial ERG recordings on patients to monitor evidence of progression or improvement would best be performed at approximately the same time of day.

1-7

THE FOCAL (FOVEAL) CONE ERG

The full-field, light-evoked ERG is a mass response to diffuse illumination of the retina and is abnormal in amplitude only when large areas of the retina are functionally impaired. The total number of cones in the central macula (a region approximately 5 mm or 20° in diameter) represents approximately only 9% of the total retinal cone population, whereas the total number of cones in the human fovea (about 1.5 mm in diameter) is only 1.5–2.0% of the entire cone population. It is therefore understandable why localized foveal lesions have little or no effect on full-field cone ERG responses.

A focal foveal cone ERG can be recorded, with the assistance of a computer, to average numerous successive responses. The procedure usually entails the use of an optical system that projects a small stimulus to a localized portion of the retina. Alternatively, a phase-alternating bar-pattern stimulus can also been used[89] (*see Section 1-8*). In any event, computer-averaged data appreciably enhance signal detection in the presence of background noise. One difficulty in recording foveal cone ERG responses with a small-field light stimulus is that light from the focal region can scatter to neighboring receptors and evoke a response that originates outside the fovea. An effective means of testing that addresses the issue of scattered light utilizes a hand-held, two-channel, Maxwellian-view ophthalmoscopic stimulator. This instrument employs a 3° white flickering (42-Hz) light stimulus, which can be focused on the fovea, and a 10° white annular surround light. The surround illumination minimizes the effect of stray light from the stimulus and provides sufficient illumination to enable continuous direct visualization of the fovea, which is necessary during the computer averaging of numerous small-amplitude responses (0.18–0.55 μV).[90,91]

Since the traditional full-field ERG recordings remain normal in the presence of localized macular disease, a reliable clinical measurement of isolated foveal cone activity demonstrated by electrophysiologic techniques would have practical diagnostic value in cases with equivocal macular findings. In conjunction with other tests of macular function, measurement of foveal cone activity would also serve as a means for objectively following the progression of macular disease. It is, however, vital to establish age-related normative data, since focal foveal cone ERG amplitudes, like full-field rod and cone ERG amplitudes, decrease with age.[91] The sensitivity of the measurement is currently such that amplitude reductions can be reliably determined in patients—for example, those with retinitis pigmentosa or juvenile macular degeneration—only at acuity levels of 20/30 to 20/40 or worse.[92] However, since normal foveal ERG amplitudes are found in strabismic amblyopia and optic atrophy,[93] this test is potentially useful for assessing patients with reduced visual acuity, particularly to levels below 20/40, that results from degeneration of foveal cones. In some eyes, foveal cone ERGs can be reduced in patients with macular disease who retain near-normal Snellen acuity. This finding may have important prognostic implications for future loss of central acuity.

Preliminary data suggest that focal foveal ERG measurements in the contralateral eyes of patients with idiopathic macular holes can be of value in predicting which of these eyes will subsequently develop a full-thickness macular hole, at least during a mean follow-up period of approximately 3 years.[94] Nevertheless, some patients with small-diameter macular holes, in addition to a high percentage of other patients with cystoid macular edema, may show normal focal ERG amplitudes.[95,96] However, patients with idiopathic central serous chorioretinopathy show reductions in local macular ERG amplitude responses with prolonged implicit times for the a-wave, b-wave, and oscillatory potentials. A greater reduction in b-wave amplitude compared to the a-wave amplitude, as well as an even greater and more prolonged reduction in amplitude of the oscillatory potentials, suggests that central serous chorioretinopathy may involve functional disturbances within the inner retinal layers in addition to the photoreceptors.[97]

1-8

THE PATTERN ELECTRORETINOGRAM (PERG)

The pattern electroretinogram (PERG) measures a retinal response to a phase-reversing pattern stimulus that maintains a constant overall mean luminance. The patterned stimulus is usually a bar grating or checkerboard configuration displayed on a television monitor in such a fashion that the bright bars or checks become dim while the dark patterns synchronously become bright.[89,98,99] When the pattern is reversed once or twice per second, a transient PERG is obtained and individual waveform components can be identified. When the pattern is reversed more than about 10 times per second, the PERG obtained resembles a sine wave and is termed a "steady-state response" (Figure 1-14). The amplitude of the PERG is small (2–4 μV). Therefore, signal averaging techniques are employed to increase the signal-to-noise ratio. Typically, 100–250 responses are averaged. An artifact rejection system must be used to eliminate artifact potentials due to eye blinks, extraocular movements, or head movements. High-contrast patterns of large elements or patterns presented at a high mean luminance are more likely to elicit recordings that are due to local luminance changes rather than to contrast or pattern-specific responses.[100,101] The main characteristics of a pattern stimulus include overall screen brightness (mean luminance), brightness contrast of neighboring bars or checks (percent contrast), rate of pattern reversal (temporal frequency), and bar or check size (spatial frequency).[100,102-104] In a description of results from PERG measurements, it is necessary to monitor and carefully stipulate these stimulus variables, as abnormal recordings in patients with various ocular disorders may be obtained with one set of stimulus parameters, but not under different stimulus conditions.

For reliability of measurement, the patient's refractive error must be corrected during testing so that the stimulus is focused on the retina[105-107] and neither

A

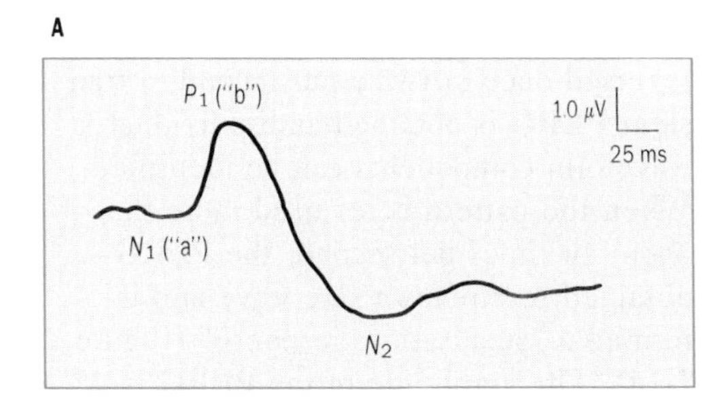

B

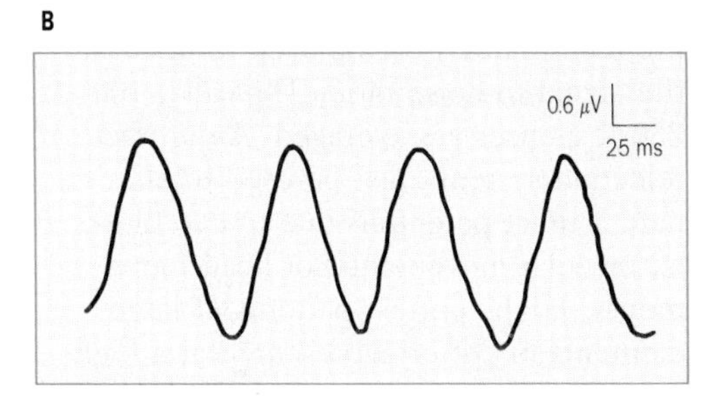

Figure 1-14 *Waveforms obtained with pattern-elicited ERG recordings. Both (A) transient and (B) steady-state responses are illustrated. P = positive; N = negative. The initial negativity (N_1) has also been labeled the "a-wave," and the large positive response (P_1) the "b-wave."* *Modified by permission from Rimmer S, Katz B: The pattern electroretinogram: technical aspects and clinical significance.* J Clin Neurophysiol *1989;6:85–99.*

the recording electrode nor any other factors should interfere with pattern clarity. Stable fixation on the fovea is also imperative. Thus, a recording electrode such as the DTL thread, which consists of silver-impregnated nylon fibers, or a gold-foil electrode is generally used since such electrodes are placed either under or over the lower lid so as not to obscure the view of the pattern stimulus. Care should be taken to place the reference electrode in an area such as the ipsilateral temple, where volume-conducted visually evoked cortical potentials will not contaminate the PERG signal. A discussion of various recording factors that can influence the PERG waveform is available in a review by Rimmer and Katz.[100]

From original work of Maffei and Fiorentini,[108] the PERG was thought to be generated within the retina, proximal to the source of the flash ERG b-wave, most probably within ganglion cells. After transection of the cat optic nerve, flash or flicker ERG recordings remained unaltered, while the pattern-evoked ERG became nondetectable over a period of 4 months. This loss of PERG followed the same time course as ganglion cell degeneration observed histologically. Similar results were subsequently observed in the monkey eye.[109]

The clinical value of the PERG for diagnosis and monitoring of disease has been assessed for such disorders as glaucoma,[110-115] optic neuritis,[116-119] optic nerve atrophy,[120-122] and amblyopia.[98,123,124] Although ganglion cells are the likely generators of the PERG response, normally functioning retinal cells distal to the ganglion cells are required to generate a normal PERG.

Thus, abnormal PERG responses can be recorded in patients with macular degenerations that affect photoreceptor cells. There is evidence that the PERG consists most likely of two separate components. The early positive wave (P_{50} or P_1) is considered to be associated with luminance aspects of the stimulus, while the subsequent negative wave (N_{100} or N_2) is thought to be more closely related to contrast or pattern-specific factors of the stimulus[125,126] (see Figure 1-14). Optic nerve disease, such as retrobulbar neuritis, can selectively reduce the amplitude of the N_2 component.[121,122] The P_1 wave originates, at least partially, from those cells that generate the full-field flash-evoked ERG, while the N_2 component appears to originate primarily from ganglion cells. As noted, however, stimulus contrast, mean luminance, and size are factors in determining the luminance versus pattern-specific contributions to the PERG waveform.

Preliminary studies suggest that the PERG may be helpful in discriminating between individuals with ocular hypertension destined to develop signs of glaucoma and ocular hypertensive persons with a more favorable visual prognosis.[114,127] Weinstein et al[113] observed that the amplitude of the N_2 wave measured with high contrast and higher luminances was reduced in patients with glaucoma and in a subset of ocular hypertensive patients regarded at greater risk for subsequently developing visual field loss. Smaller-size, counterphasing checkerboard patterns were found by Bach et al[115] to demonstrate greater reductions of amplitudes in patients with early stages of chronic open-angle glaucoma than did larger check sizes. Abnormal amplitudes in the PERG have also been noted in diabetic patients with background retinopathy.[128,129]

1-9

THE ERG IN RETINAL DISORDERS

The various retinal and choroidal diseases or disorders that can affect the ERG will be referred to under the following headings: photoreceptor dystrophies, stationary night-blinding disorders, presumed primary dystrophies of the pigment epithelium, chorioretinal dystrophies, Bruch's membrane disorder, hereditary vitreoretinal disorders, inflammatory conditions, circulatory deficiencies, toxic retinopathy, nutritional (vitamin A) deficiency, optic nerve and ganglion cell disease, opaque media (lens-vitreous), diabetic retinopathy, and miscellaneous conditions.

Figure 1-15 illustrates the patterns of abnormal ERG responses in photopic and scotopic conditions. The terms "negative-negative" and "negative-positive" were used by Henkes[130] for relatively lower luminance recordings. Both are characterized by a more prominent a-wave amplitude compared to the b-wave. In the negative-negative response the b-wave amplitude is decreased, whereas in the negative-positive response the b-wave amplitude remains normal while the a-wave amplitude is increased. It is likely that the negative-positive response can occur as a variant of normal, although Henkes described its occurrence in some patients with central retinal vein occlusion. Van Lith[131] argued effectively that

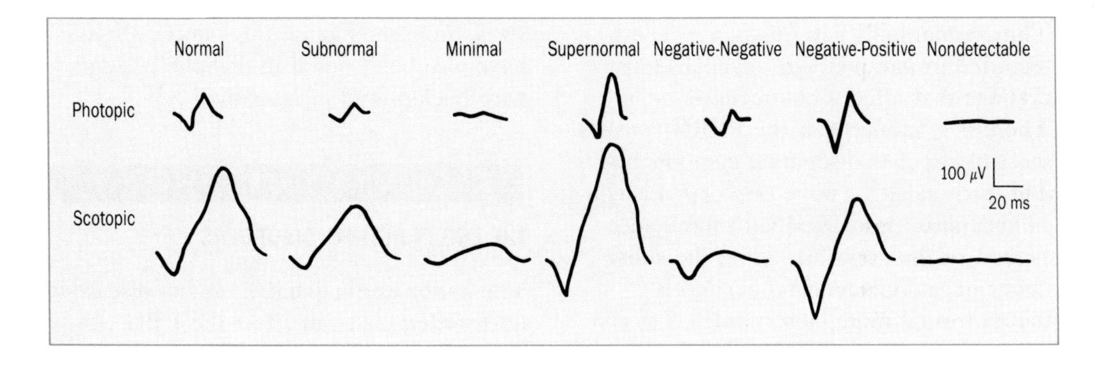

Figure 1-15 *Patterns of normal and abnormal ERG responses under photopic and scotopic conditions.*

qualitative descriptions or labels of ERG waveforms as "subnormal" or "nondetectable" are less desirable than quantitative determinations that document a percent reduction in ERG amplitudes. Van Lith suggested expressing amplitude reduction as a percentage of the mode and lower limit. The mode is used instead of the mean since in most laboratories there is an asymmetric frequency distribution of normal ERG amplitude values. Nevertheless, van Lith determined the lower limit by obtaining a standard deviation value of 2 in reference to the mode. Reductions in ERG amplitudes were then expressed as a percentage of the mode and the lower limit. In instances of uniocular disease involvement, van Lith recommended using the uninvolved eye as a control when its amplitude is greater than the mode. Regardless of whether the mode, median, or lower limit (however determined) is used, the fundamental issue is that describing an ERG result solely as "subnormal" is inexact and in need of quantitative documentation.

With reference to "absent" ERG recordings, it is advocated that a mean background or noise level be determined so that terms such as "nondetectable," "nonrecordable," or "extinguished" can be stated in a more precise, quantitative manner. Thus, the term "nondetectable" should be expressed with reference to a specific background level under a specific set of stimulus and recording conditions. Of necessity, this monograph will use such nonquantitative terms as "subnormal" and "nondetectable" since reports on various retinal diseases in the literature, which will now be addressed, rarely provide quantitative definitions of these terms.

1-10

PHOTORECEPTOR DYSTROPHIES OR DYSFUNCTIONS

1-10-1 Rod-Cone (Retinitis Pigmentosa)

First described, as seen through the ophthalmoscope, by van Trigt in 1853 at the Donders clinic in Utrecht, the Netherlands, this group of diseases was first named retinitis pigmentosa in 1857 by Donders. Although patients characteristically show a conspicuous pigmentary degeneration of the retina, ERG changes can precede ophthalmoscopically visible fundus disease in all genetic patterns. In most instances, patients with either the autosomal recessive or the X-linked recessive form have nondetectable or very reduced responses at an early age. Computer averaging of several responses and the optimal use of frequency filters can aid in the detection of very small responses previously obscured within the noise level of the recording procedure.[39] The ultimate reliability and thus value of obtaining these ERG recordings (which can be less than 1 μV) for patient management and/or follow-up have yet to be conclusively determined. When responses are obtainable, from standard recording procedures, they are those from a high-luminance stimulus and include either minimal or appreciably subnormal a- and b-waves. In one study of X-linked recessive patients, 71% showed nondetectable cone and rod amplitudes using conventional full-field recording procedures.[132] In those with recordable responses, approximately 50% showed recordings in which the responses from rods were more favorably retained than those from cones, while in an approximately equal percentage, cone amplitudes were relatively less severely affected than rod amplitudes.

The autosomal dominant varieties of retinitis pigmentosa not infrequently have a later onset of symptoms and a less rapid progression. When ERG recordings are measurable, in at least some subtypes, they are either predominantly or exclusively from remaining cone receptor cells (Figure 1-16). Nevertheless, in earlier stages, some patients with autosomal dominant disease will manifest subnormal, but recordable cone and rod ERG responses. However, several subtypes of autosomal dominant inheritance are seen with varying degrees of severity.[133] On the basis of rod sensitivity relative to cone sensitivity, determined by perimetric dark-adapted absolute threshold measurements to a short- and long-wavelength stimulus, at least two subtypes of autosomal dominant retinitis pigmentosa are identified.[134,135] Electroretinogram recordings done in conjunction with these psychophysical measurements of absolute threshold may provide supplemental information to better facilitate categorization of patients with dominantly inherited retinitis pigmentosa.[136] Autosomal recessive and isolated (or simplex) retinitis pigmentosa patients can similarly be subdivided by perimetric dark-adapted absolute threshold measurements.[135,137]

In all three genetic forms, cone and rod responses (when present) can be prolonged in implicit time.[138,139] However, a prolonged dark-adapted b-wave implicit time is not necessarily a characteristic property of the ERG in retinitis pigmentosa patients.[61,140,141] Prolonged cone b-wave implicit times to a single-flash stimulus are likely to be apparent, particularly when rod function has become appreciably impaired. In this regard, a rod influence on cone b-wave implicit time has been postulated.[58] Of interest, a 30-Hz flicker response may also show a prolongation of cone implicit time. Sandberg et al[142] also demonstrated a prolongation in the dark-adapted cone a-wave implicit time in retinitis pigmentosa patients, a

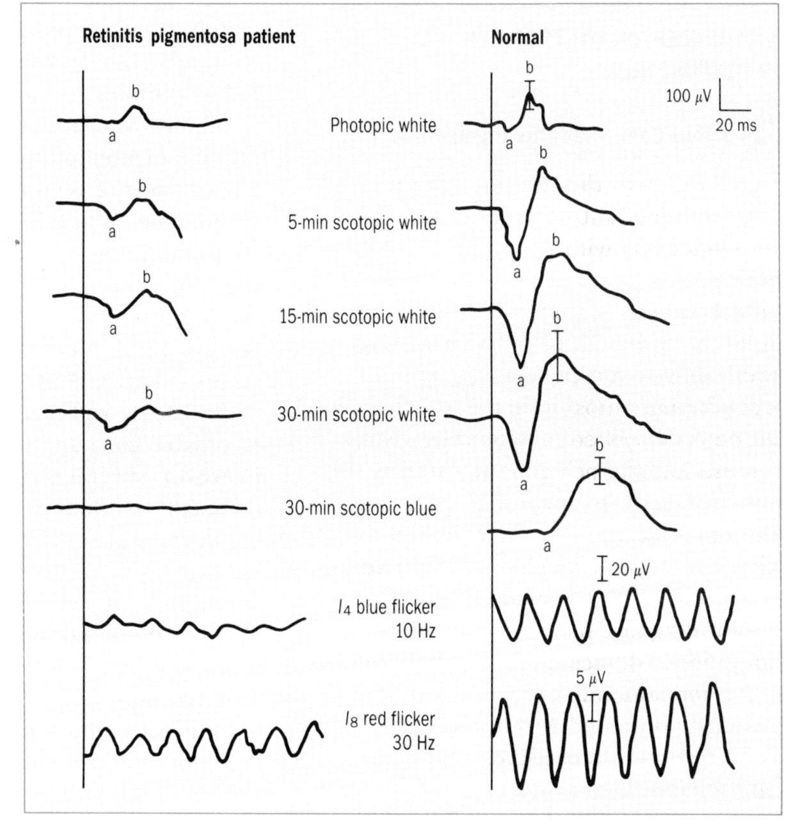

Figure 1-16 *ERG recording from a patient with autosomal dominant retinitis pigmentosa. The remaining responses under both photopic and scotopic levels of adaptation are primarily from the cone system.*

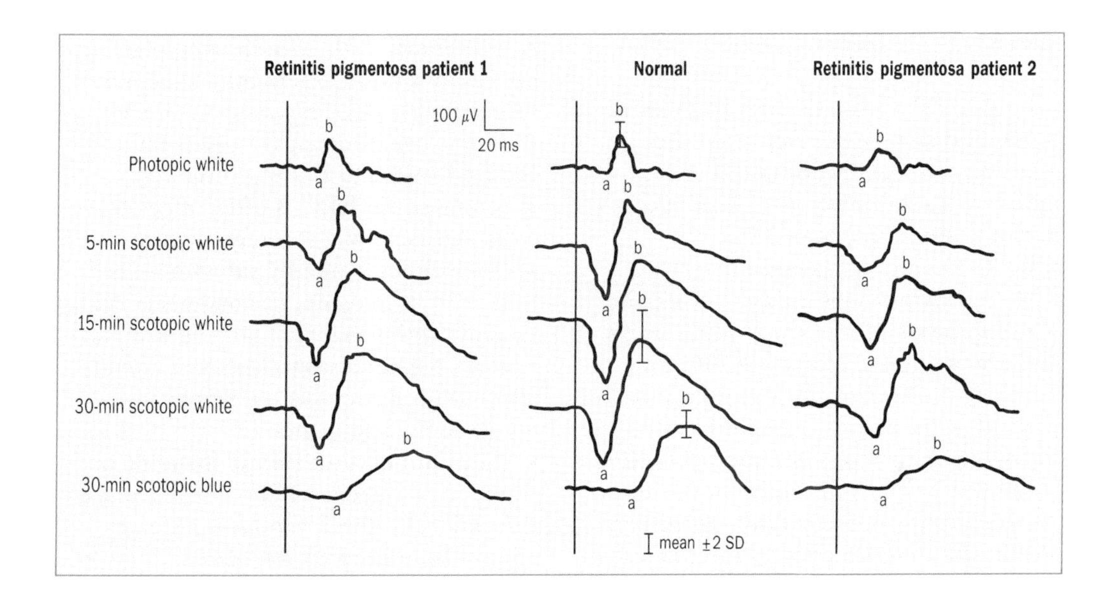

Figure 1-17 *Two patients from the same family with a "delimited" form of autosomal dominant retinitis pigmentosa demonstrating a substantial ERG response. Cone and rod b-wave implicit times are normal.*

finding that the authors interpreted as consistent with shortened cone outer segments found in these patients. In the early stages of autosomal dominant retinitis pigmentosa with reduced penetrance, a prolongation of cone b-wave implicit time can be noted even in the presence of normal cone amplitudes.[139] Patients who have an early-stage autosomal dominant retinitis pigmentosa with complete penetrance can show cone b-wave responses that are initially normal in both amplitude and implicit time.[143] Berson and Kanters[144] noted, at least in some families with autosomal recessive retinitis pigmentosa, that both a reduced cone b-wave amplitude and a prolonged implicit time can precede detectable abnormalities in the rod system. In some instances, patients with a "delimited" form of retinitis pigmentosa have surprisingly substantial ERG responses under both light- and dark-adapted conditions (Figure 1-17).

These patients seem to have a more favorable prognosis compared to those with more extensive reductions in ERG amplitudes. Although seen relatively infrequently, one report indicated that they may comprise as many as 17% of patients with retinitis pigmentosa.[140]

In addition to its value in diagnosis, ERG testing provides a means of monitoring progression of this group of disorders and evaluating the efficacy of future therapeutic trials. Moreover, in at least some patients with retinitis pigmentosa, a reduction in early receptor potential (ERP) amplitude has been recorded in the absence of ophthalmoscopically visible bone-spicule–like fundus pigmentation or arteriolar narrowing.[145] Theoretically, reductions in ERP amplitudes might be more apparent than reductions in ERG amplitudes in patients with very early stages of pigmentary retinal degeneration. This notion awaits experimental verification.

The carriers of X-linked recessive retinitis pigmentosa often show fundus changes. These include a golden metallic-luster appearance within the posterior pole (referred to as a tapetal-like reflex), local areas of peripheral pigmentary changes (including pigment atrophy and bone-spicule–like pigmentary clumping), fundus changes seen in high myopia, and, on occasion, extensive and diffuse pigmentary degeneration, as seen in affected male patients.[146] Approximately 86–90% of carriers can be identified by reductions in ERG amplitudes.[146] Bright stimulus flashes have been demonstrated as most effective for carrier identification by ERG criteria.[147]

Syndromes with an associated pigment dystrophy similar to retinitis pigmentosa have variable degrees of abnormal electrophysiologic functions. In Refsum's, Bardet-Biedl, and Bassen-Kornzweig syndromes, the ERG is often affected early and generally severely, with responses either markedly subnormal, minimally recordable, or nondetectable. Patients with congenital hearing loss and retinitis pigmentosa (Usher's syndrome) also have considerable reductions in ERG amplitudes. The degree of functional visual impairment can vary, depending on the type of Usher's syndrome.[148] Detailed discussions of these various syndromes are available in other references.[149-152] In Bassen-Kornzweig syndrome, abnormal ERG recordings have been reported in children whose fundus was still normal.[153] Of interest is the observation that patients with Refsum's disease and Bassen-Kornzweig syndrome tend not to show the delays in ERG b-wave implicit time that can be noted in patients with retinitis pigmentosa, for whom biochemical defects have not yet been identified.[154,155] In Bassen-Kornzweig syndrome, in which patients show low to minimal levels of serum vitamin A, parenteral water-soluble vitamin A supplements increase the electrical activity from both rods and cones in some cases with less severe fundus changes.[156-158] Early treatment with vitamins A and E may retard the development and progression of retinal photoreceptor cell degeneration in this disorder.[159,160]

In the mucopolysaccharidoses—particularly, Hurler's (MPS I-H), Sanfilippo's (MPS III), and Scheie's (MPS I-S) syndromes—an associated pigmentary retinopathy causes ERG amplitudes to vary from subnormal to nondetectable. Morquio's and Maroteaux-Lamy syndromes are not generally associated with pigmentary retinopathies and the ERG amplitudes are usually normal. Patients with Hunter's syndrome may have either normal or abnormal ERG recordings, depending on the presence or absence of a pigmentary retinopathy. More detailed discussions of ERG findings in the mucopolysaccharidoses, in addition to their classification and characteristics, are available elsewhere.[161,162]

Electroretinogram recordings may also be of value in helping to confirm the diagnosis or document the severity of retinal degeneration in children with various forms of lipopigment storage disorders (so-called neuronal ceroid lipofuscinosis or Batten's disease), in which appreciable reduction in ERG cone and rod amplitudes can be found in the infantile (Hagberg-Santavuori syndrome[163]), late infantile (Jansky-Bielschowsky disease), and juvenile (Spielmeyer-Vogt disease[164]) forms. Initially, the ERG can present with a selective reduction of the b-wave amplitude under both light- and dark-adapted conditions.[165] A selective reduction in the scotopic ERG b-wave amplitude has been noted in at least some heterozygotes for the juvenile form of ceroid lipofuscinosis.[166]

The variants of classic retinitis pigmentosa also show diverse patterns both morphologically and electrophysiologically. Previously, the term "retinitis pigmentosa sine pigmento" was used to describe patients who clinically showed a normal retina despite documented diffuse abnormalities of photoreceptor cell function. This term is now appropriately regarded as confusing since it implies a distinct disorder rather than either a less extensive stage or a variant in expression of retinitis pigmentosa. Patients previously described as having retinitis punctata albescens have a progressive retinal degeneration with night blindness and peripheral visual field loss. The retina characteristically shows generally diffuse, white, punctate lesions located deep within the retina. Not infrequently, pigment clumping with a bone-spicule–like configuration is also present.[167] The sine pigmento and retinitis punctata albescens groups generally have significantly diminished or absent ERG responses. When present, responses have prolonged b-wave implicit times. Either retinitis pigmentosa sine pigmento or retinitis punctata albescens may occur in a patient when other family members show the typical retinitis pigmentosa fundus appearance.

Unilateral retinitis pigmentosa refers to a pigmentary retinal degeneration associated with minimal or nondetectable ERG responses within the affected eye, while the other eye is clinically and functionally normal. Henkes[168] noted the importance of obtaining electrophysiologic studies on the apparently normal eye to assure that it is also functionally normal. Carr and Siegel[169] emphasized the probable secondary inflammatory, traumatic, and vascular causes in most cases of presumed unilateral retinitis pigmentosa. These authors presented cases of ophthalmic artery occlusion that simulated a unilateral retinitis pigmentosa. Kandori et al[170] also stressed the importance of ophthalmic artery occlusion as causing a pigmentary retinopathy and reported two cases with a secondary, apparently vascular cause of pigmentary retinopathy. Kolb and Galloway[171] noted the absence of a positive family history in these cases and, like Henkes,[168] reported subnormal electro-oculograms (EOGs) in the normal eyes of patients with clinically apparent unilateral retinitis pigmentosa. Thus, most authors believe that true unilateral retinitis pigmentosa is a rare entity because of (1) the absence of a family history typical of the other classic forms of retinitis pigmentosa, (2) known traumatic, vascular, and inflammatory causes in some cases of unilateral pigmentary retinopathy, and (3) abnormal ERG and/or EOG values in the apparently normal fellow eye of some patients with unilateral retinitis pigmentosa. Perhaps a more appropriate term for this disorder would be "unilateral retinal pigmentary degeneration."[152]

Sector retinitis pigmentosa is infrequently encountered. The fundus has local areas of constricted retinal vessels and bone-spicule pigment clumping around or within the quadrant of these vessels. The disease is characteristically bilateral and symmetric, with the abnormality most frequently residing within the inferior and nasal quadrants, producing corresponding defects in the superior and superior-temporal visual fields. The ERG is generally subnormal, similar to that found in the earlier stages of some forms of autosomal dominant retinitis pigmentosa. Contrasting, however, are the normal cone and rod implicit times in the sector form[154,155] (Figure 1-18). The alteration in the ERG amplitude is, in most cases, proportional to the extent of ophthalmoscopically apparent retinal involvement. This disease is most frequently inherited as an autosomal dominant trait and is slowly progressive. Although the disease is clinically limited to a sector of the retina, functional impairment can frequently be demonstrated by psychophysically determined visual thresholds even in clinically normal-appearing quadrants.[172] Thus, this disorder would better be termed asymmetric rather than sector if consideration is given to functional impairment rather than to the clinical fundus appearance.

Leber's congenital amaurosis is a heterogeneous group of primarily autosomal recessively inherited disorders that represent a congenital form of diffuse photoreceptor cell disease. Patients, in general, show retinal changes phenotypically similar to those seen in retinitis pigmentosa. This entity is not rare and was found to account for 10% of all blindness in Sweden and 18% of all blindness in children in Holland. Although some degree of pigmentary change is apparent in these patients, even at an early age, it is important to consider this diagnosis in all infants whose behavior and physical findings—including nystagmus, an oculodigital sign, and photoaversion—suggest poor vision even in those whose retinal examination appears clinically normal. Once poor vision is suspected, an ERG recording that shows either a nondetectable or a markedly reduced amplitude can prevent needless and often costly neurologic and radiologic examinations looking for a "central" cause of blindness. This recording

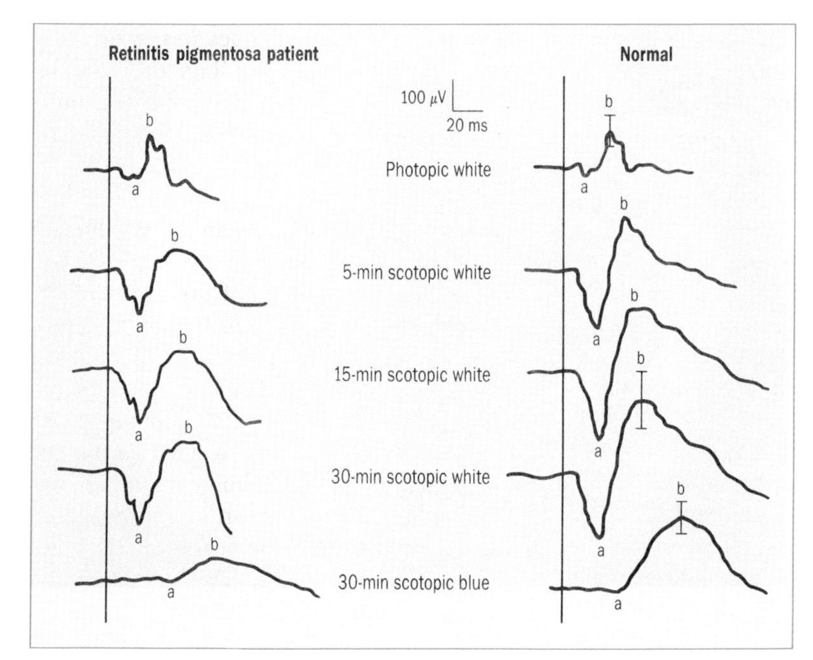

Figure 1-18 *ERG responses from a patient with sector retinitis pigmentosa. As in patients with "delimited" retinitis pigmentosa, cone and rod b-wave implicit times are normal.*

should be obtained only with topical anesthetic since general anesthesia is rarely justified for an ERG in these infants or even in most young children. Other ocular findings may include keratoconus and/or cataract. In certain patients, an erroneous diagnosis of a macular coloboma or congenital toxoplasmosis may be made because of the presence of a large atrophic macular lesion unless an ERG recording is obtained.[173] Some children have associated neurologic or psychiatric disturbances, including mental retardation.

1-10-2 Cone and Cone-Rod

The acquired cone or cone-rod dystrophies represent a heterogeneous group of disorders that are inherited as an autosomal dominant, autosomal recessive, or X-linked recessive trait. The dominant forms most frequently manifest a "bull's-eye" pattern within the macula with either normal or slightly attenuated retinal vessels (Figure 1-19). A temporal optic disc pallor may also be visible. The autosomal recessive forms frequently have a more diffuse central atrophy, often without a clear bull's-eye appearance. Both groups characteristically have loss of central acuity, central scotomas, and defective color vision. Photoaversion and, not infrequently, nystagmus are also found. As might be expected, the single-flash photopic ERG is reduced in amplitude. The flicker ERG response is also abnormal both in amplitude and in fusion frequency (Figure 1-20). These patients can show a nondetectable recording at stimulus frequencies

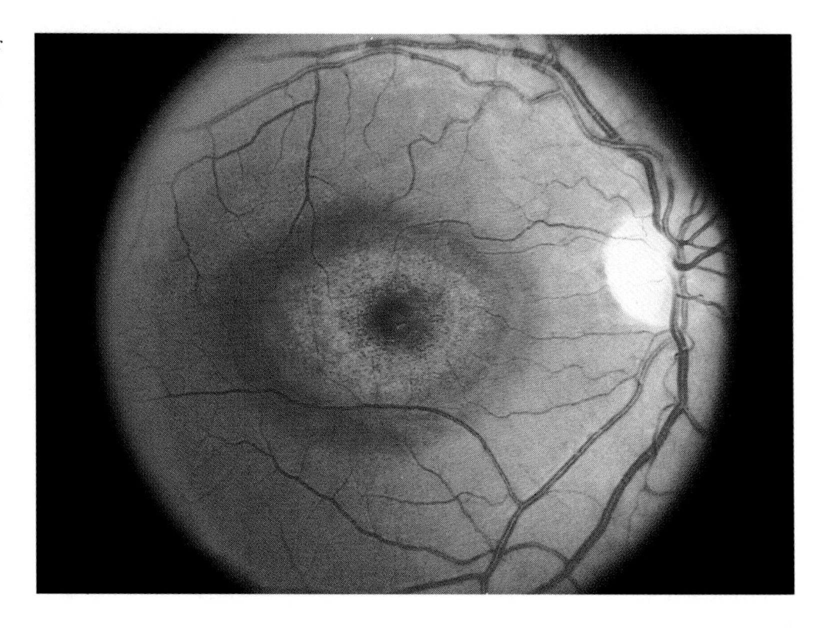

Figure 1-19 *Characteristic bull's-eye macular lesion seen in a patient with cone dystrophy.*

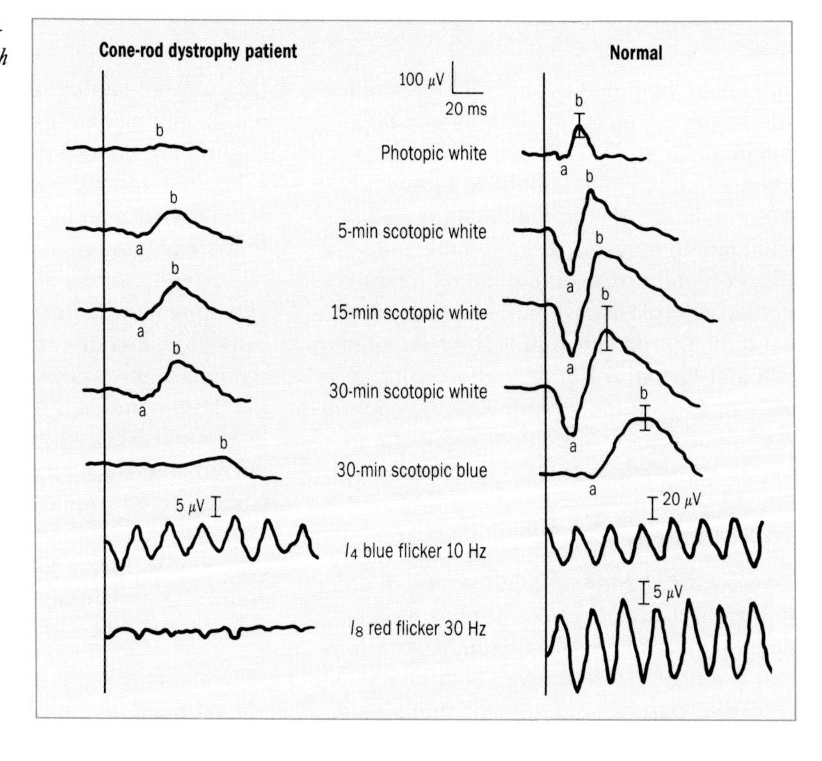

Figure 1-20 *ERG recording from a patient with cone-rod dystrophy. Note that cone responses are barely detectable, while rod responses, although appreciably impaired, are considerably more preserved than cone responses.*

of 30 Hz, while the range of normal values is approximately 55–70 Hz.

In patients with cone dystrophy, rod responses are entirely normal or only minimally affected in the early stages. With time, rod as well as cone responses can show additional impairment.[174] Some autosomal recessive and, less frequently, autosomal dominant forms can, however, manifest appreciable rod ERG amplitude reduction even at the time of diagnosis. The reduced rod ERG amplitudes are associated with a generally normal, but occasionally prolonged, implicit time. Such patients are best described as having a cone-rod dystrophy. At least some patients previously described as having inverse retinitis pigmentosa were likely to have had a form of this disease. Cone-rod dystrophy patients infrequently complain of poor night vision, but can manifest contracted peripheral fields or ring scotomas.[175] Usually, the fundus examination then shows retinal vessel attenuation and midperipheral as well as peripheral pigment clumping in addition to a macular lesion (Figure 1-21). Optic disc atrophy and disc vessel telangiectasia have also been reported in these patients.[176] In isolated instances, patients have been reported in whom peripheral cone function was apparently more impaired than central cone function.[174,177] However, other patients have a primarily central cone dystrophy since their full-field ERG cone function is normal.[178]

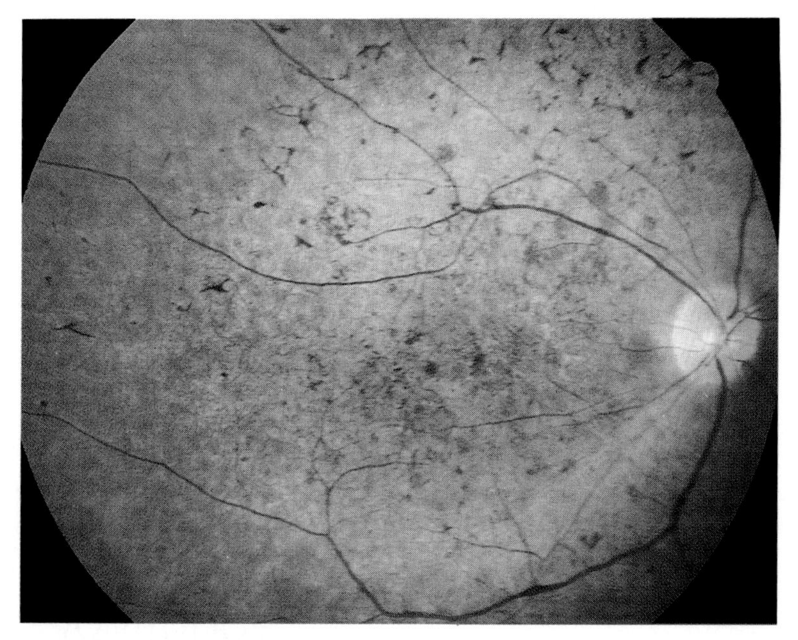

Figure 1-21 *Extensive pigmentary degenerative changes in a patient with cone-rod dystrophy. Some degree of phenotypic similarity to bone-spicule–like pigment changes seen in patients with retinitis pigmentosa is apparent.*

Congenital achromats are classified as typical (rod) and atypical (cone) monochromats. Although described in this section with photoreceptor dystrophies, such patients have a nonprogressive visual disorder and therefore would best be described as having a cone dysfunction rather than a dystrophy, which implies a progressive deterioration of function. The term monochromat reflects these patients' perception of all colors as shades of gray. The rod monochromat, who can show either complete or incomplete involvement, is the form most frequently seen clinically, although its prevalence is only 3 in 100 000 people. These patients are similar to cases of acquired cone dystrophies in having photoaversion, nystagmus, decreased visual acuity, and poor color perception. Generally, they have a normal fundus or only minimal, nonspecific pigment changes within the macula. The photopic electrical activity of the ERG is generally nondetectable, while scotopic responses remain either normal or minimally subnormal (Figure 1-22).

The incomplete rod monochromat is similar in characteristics to the complete form except for less extensive involvement. Photoaversion and nystagmus are generally less apparent. Although visual acuity is usually in the range of 20/60 to 20/80, it can be as good as 20/40. ERG photopic responses are markedly reduced in amplitude, although, in certain instances, less extensively than in the complete monochromat. The incomplete and complete forms probably represent modifi-

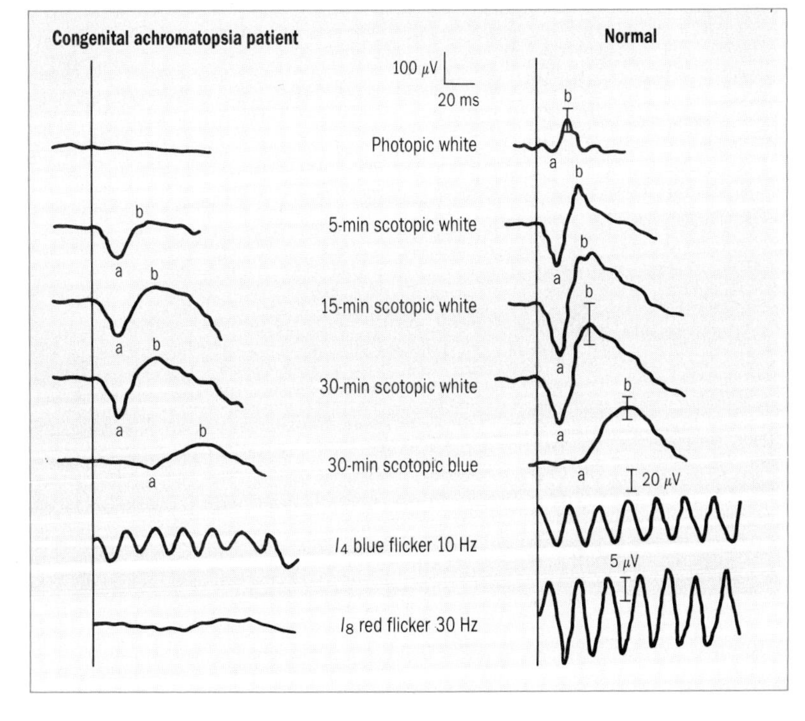

Figure 1-22 *ERG recording from a patient with congenital achromatopsia. Cone responses are nondetectable, while rod amplitudes are at the lower range of normal.*

cations of one genotype, since in some families the members may alternately manifest the complete or the incomplete form.

A unique group of patients with incomplete achromatopsia are those in whom the blue cones are minimally involved or not involved. These blue-cone monochromats have abnormal ERG amplitude responses similar to those of complete achromats, since blue cones contribute only negligibly to the full-field ERG. The condition is inherited as an X-linked recessive trait.

The atypical, or cone, monochromat is characterized by normal acuity and the absence of nystagmus and photoaversion. In this rare condition (affecting approximately 1 in 100000 to 1 in 10000000 people), cone outer segment function is probably normal. There appears to be a transmission defect involving cones that leads to a defect in chromatic perception. These patients maintain a normal quantity of photoreceptors. Generally, patients have a normal ERG to a white light stimulus, attesting to the notion of a postreceptoral, or transmission, defect.[179]

1-11

STATIONARY NIGHT-BLINDING DISORDERS

1-11-1 Congenital Stationary Night Blindness With Myopia and Normal-Appearing Retina

Autosomal recessive, X-linked recessive, and autosomal dominant forms of congenital stationary night blindness are all found. One subtype with a dominant mode of inheritance is known as the Nougaret type, named for Jean Nougaret, the first affected member in a French genealogy with autosomal dominant congenital stationary night blindness traced through 11 generations. The X-linked and autosomal recessive patients are generally myopic (−3.5 to −14.5 D) and have a history of poor night vision. The visual acuity of patients with the X-linked form is between 20/40 and 20/200. Those patients with the autosomal recessive form who have a decrease in visual acuity (to between 20/40 and 20/80) are usually also myopic. However, some of the reported patients with both the X-linked and the autosomal recessive forms have normal visual acuity and are not myopic. Further, in some instances, patients may not complain of night blindness.[180] In this circumstance, a definitive diagnosis can be made only by characteristic b-wave amplitude changes on full-field ERG recordings. Patients with visual loss frequently manifest nystagmus. Patients with the autosomal dominant form are generally not myopic and tend to have normal visual acuity. Traditionally, findings of the fundus examination in all groups are normal, although the optic disc has been described as tilted, pale, or dysplastic.[181]

The scotopic ERG responses are predictably those most notably affected. However, photopic (cone) responses, although reported as normal in amplitude in some patients, not infrequently show a selective reduction in b-wave amplitude.[182-184] Although some investigators have emphasized normal photopic b-wave implicit times,[183] others have reported patients with prolonged photopic b-wave implicit times.[182,184] In X-linked and autosomal recessive forms, a negative ERG pattern also occurs, with the scotopic a-wave larger in amplitude than the reduced scotopic b-wave (Figure 1-23). An important observation is the approximate equality of b-wave implicit times under both photopic and scotopic conditions to a similar intensity stimulus in many patients with stationary night blindness of either X-linked or autosomal recessive inheritance. Normally, the scotopic b-wave implicit time is at least twice as long as that of the photopic b-wave. Thus, in this pattern (Schubert-Bornschein type), a negative ERG is found with an essentially normal scotopic a-wave noted along with a reduced b-wave that consists of a predominantly cone component, since, in these patients, the rods contribute minimally if at all to the b_1-wave amplitude or the formation of a b_2 component. However, in patients with high degrees of myopia, a reduction in a-wave amplitude compared to normals may be apparent, although not as extensive as that noted for the b-wave amplitude.

In a second pattern (Riggs type) noted in some cases with congenital stationary night blindness, the scotopic b-wave exceeds the a-wave, but is still reduced in amplitude. The scotopic a-wave either obtains a normal amplitude or, less frequently, remains slightly subnormal. Although the photopic component appears

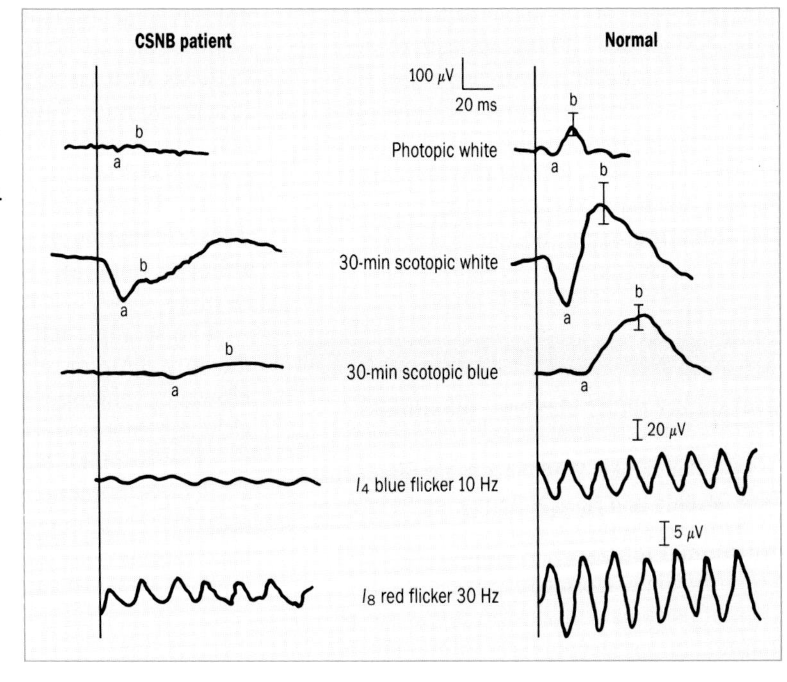

Figure 1-23 *Selective b-wave reduction to high-luminance stimulus as characteristically seen in a patient with congenital stationary night blindness (CSNB). This negative ERG pattern is found in stationary night blindness patients who are referred to as having a Schubert-Bornschein type.*

normal, there is a defective development of the rod contribution to the dark-adapted (scotopic) b-wave. The result is a reduction in b_1 amplitude and a failure of normal b_2 formation. The photopic and scotopic implicit times, unlike those found with the Schubert-Bornschein pattern, are more similar to the normal time relationships anticipated under photopic and scotopic conditions.

Patients with congenital stationary night blindness appear to have a normal in situ amount or density of rod visual pigment and, in the Schubert-Bornschein type, a normal a-wave. Further, the rod photopigment regeneration kinetics are entirely normal. In such cases, the photoreceptor cells are therefore intact and the primary defect probably does not reside within these cells but rather in the transmission of their visually evoked signal. Carr et al[185] studied a patient with autosomal recessive and another with autosomal dominant congenital stationary night blindness. By fundus reflectometry studies, both patients manifested a normal concentration and regeneration of rhodopsin. In the autosomal recessive form, the scotopic a-wave was normal while the scotopic b-wave was reduced in amplitude. A normal EOG light rise was seen. In the autosomal dominant form, the EOG light rise was abnormal and all ERG responses, both photopic and scotopic a- and b-waves, were suppressed. These authors concluded that the pathology probably reflected a defect in transmission of the ERG wave potentials. For the autosomal dominant form, this defect possibly resides at the level of the photoreceptor inner or outer segment. Alternatively, a defect could reside in a breakdown of the mechanism whereby quantal absorption is converted to an electrochemical receptor potential, quite possibly from a defect in one of the light-activated enzymatic stages in the transduction process. The recessive variety, the authors speculated, possibly has a defect in, or just distal to, the bipolar cell layer.

Miyake et al[37] classified patients showing a Schubert-Bornschein ERG pattern into "complete" and "incomplete" subtypes based on psychophysically measured thresholds and ERG amplitudes. In the complete form, absolute psychophysical thresholds are mainly cone-mediated; the rod ERG responses to a dim, short-wavelength (blue) flash are absent; and the cone ERG amplitude is only moderately reduced. In the incomplete form, absolute thresholds, although elevated above normal, are rod-mediated; the rod ERG responses, although reduced in amplitude, are still measurable to a dim, short-wavelength flash; while the cone ERG is markedly subnormal. Therefore, in the incomplete type, the rod system is less affected and the cone system more impaired than in the complete form. In the incomplete type, the oscillatory potentials are more frequently recordable than in the complete type. Further, incomplete-type patients show an exaggerated increase in 30-Hz flicker ERG amplitude and a characteristic change of waveform shape (separation phenomenon) during light adaptation following 30 minutes of dark adaptation.[186] Many patients in the complete group have high or moderate myopic refractive errors, while those in the incomplete group tend to have mild myopic or hyperopic refractive errors. To date, no pedigree has convincingly shown a complete and an incomplete type in the same family. ERG recordings, in addition to psychophysical measurements, are of obvious value in distinguishing these two distinct subtypes.

Miyake and Kawase[187] reported a selective reduction in the amplitude of the oscillatory potentials in female carriers of X-linked recessive congenital stationary night blindness. This observation could be of practical value for the determination of the carrier state for females at 50% risk in families with this disorder.

Two patients with cutaneous malignant melanomas have been reported with an acquired acute onset of night blindness and ERG recordings similar to those seen in patients with congenital stationary night blindness and a selective b-wave amplitude reduction.[188,189] In one patient, the use of vincristine[188] was felt to account for the findings, since in the perfused cat eye this drug can cause a selective reduction in the ERG b-wave.[190] The second patient was not, however, receiving any drugs prior to the onset of night blindness and ERG changes.[189]

1-11-2 Oguchi's Disease

Patients with Oguchi's disease have an early history of defective night vision. The majority of the relatively rare cases of this autosomal recessively inherited disease are from Japan, although even as early as 1963 Franceschetti et al[191] noted 32 European cases. The vision and peripheral fields in these patients are generally normal in daylight; however, with dim illumination, they have a manifest defect in visual sensitivity. The photopic ERG functions are normal, while the scotopic b-wave is diminished in amplitude, producing a negative ERG pattern with high-luminance stimuli, as seen in congenital stationary night blindness. Thus, as with congenital stationary night blindness, there is an absence or marked reduction in b_2 amplitude and a reduction in b_1 amplitude. Oscillatory potentials have been reported variously as normal or reduced in amplitude,[192] while the EOG has been found to be normal.

The disease is characterized by a yellowish phosphorescent-like discoloration, most often within the peripheral retina. In most cases, this fundus discoloration returns to normal after several hours of dark adaptation (Mizuo-Nakamura phenomenon). The kinetics of the process of dark adaptation do not coincide with the return to normal fundus coloration. With prolonged periods of dark adaptation, most patients with Oguchi's disease manifest an improvement in dark-adaptation thresholds and ERG b-wave amplitudes to normal or near-normal values. Rod and cone b-wave implicit times are normal after appropriate periods of adaptation.

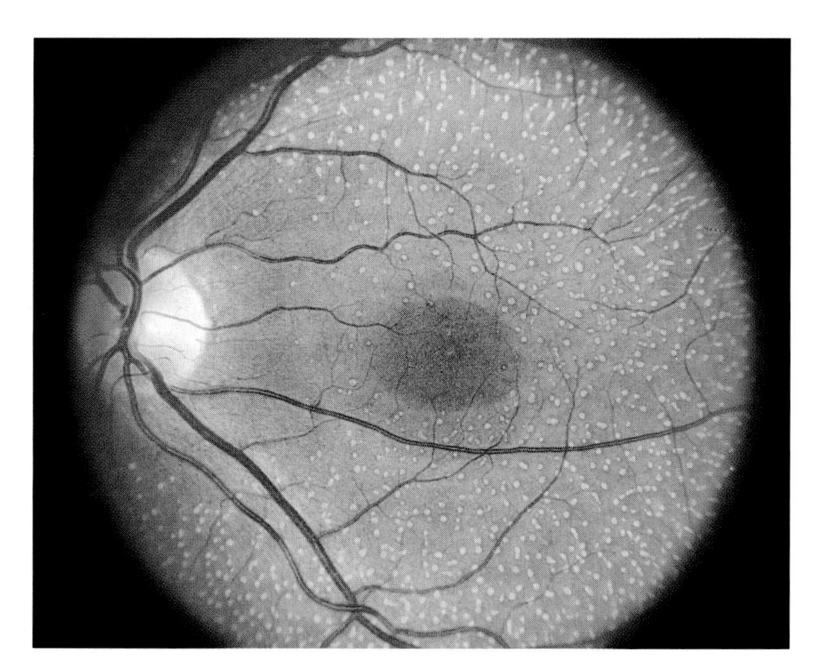

Figure 1-24 *Fundus photograph of a patient with fundus albipunctatus, showing multiple spots of unknown material scattered primarily throughout the deep retina with the exception of the fovea.*

The site of the retinal pathology and the nature of the photosensitive pigment responsible for the Mizuo-Nakamura phenomenon are not yet known. Rhodopsin kinetics, including both concentration and regeneration as measured by fundus reflectometry, are normal.[193] It seems reasonable to suggest that the normal process of dark adaptation is compromised in patients with this disorder by abnormal function of the postreceptoral cellular elements responsible for generating the ERG b-wave.[193]

1-11-3 Fundus Albipunctatus

The autosomal recessive disorder fundus albipunctatus is characterized by an early onset of nonprogressive poor night vision and the presence of numerous yellowish white spots scattered throughout the fundus with the exception of the fovea (Figure 1-24). The optic disc and retinal vessels are normal, and there is no evidence of peripheral clumping of retinal pigment. Both cone and rod adaptations follow a markedly prolonged time course, ultimately reaching normal thresholds. Correspondingly, ERG and EOG responses eventually reach normal values, but only after a prolonged period of dark adaptation prior to testing.

It seems likely that this disorder results from a defect at some stage in the cycle of visual pigment regeneration, which has been demonstrated to be prolonged.[194]

1-12

PRESUMED PRIMARY DYSTROPHIES OF THE PIGMENT EPITHELIUM

1-12-1 Stargardt's Macular Dystrophy

Stargardt's macular dystrophy (fundus flavimaculatus) was first described in 1909 by Karl Stargardt, a German ophthalmologist. Patients with this predominantly autosomal recessively inherited disorder characteristically present within the first 10–20 years of life, with a decrease in central acuity and bilateral atrophic-appearing foveal lesions. These lesions may have a beaten-metal appearance and are associated with a ring or more extensive garland of yellow-white fundus flecks within the posterior pole and, to a lesser extent, the midperipheral retina (Figure 1-25). The fleck-like lesions may be round, linear, or pisciform (fishtail-like). Impairment in central acuity is progressive, with ultimate visual acuity generally reaching 20/200 to 20/400. Initially, only subtle pigmentary mottling within the fovea may be apparent ophthalmoscopically. Even with only these minimal ophthalmoscopic changes, visual acuity may decrease to levels of 20/30 to 20/50.[195,196]

Initial investigators preferred the term "Stargardt's macular dystrophy" when referring to patients with an atrophic-appearing foveal lesion and a limited number of fundus flecks and the term "fundus flavimaculatus" (first used in 1963 by Franceschetti[197]) for patients with extensive fundus flecks with or without an atrophic foveal lesion. Subsequent studies support the notion that the above dichotomy of patients with and without extensive flecks, independent of the presence or absence of a foveal lesion, into distinct

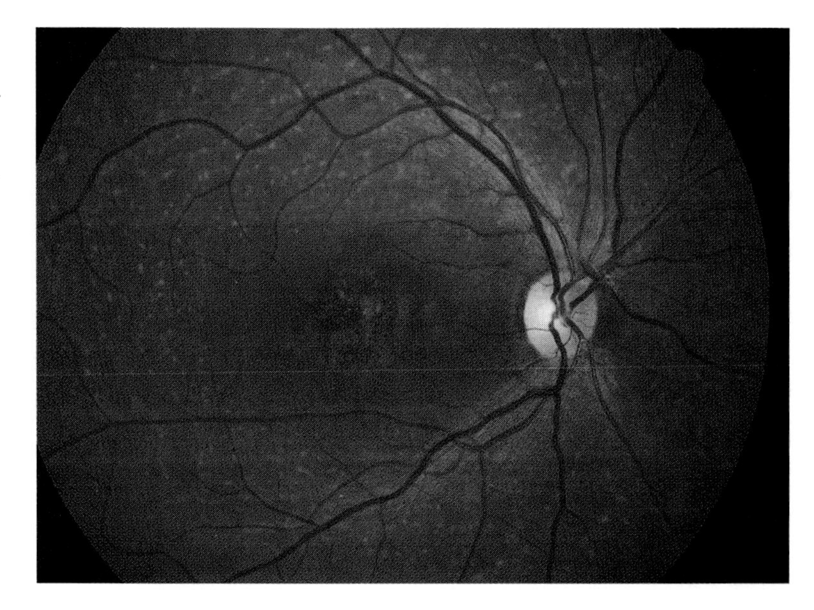

Figure 1-25 *Fundus photograph of a patient with Stargardt's macular dystrophy (fundus flavimaculatus). Note the atrophic-appearing macular lesion, with numerous fleck-like deposits throughout the retina.*

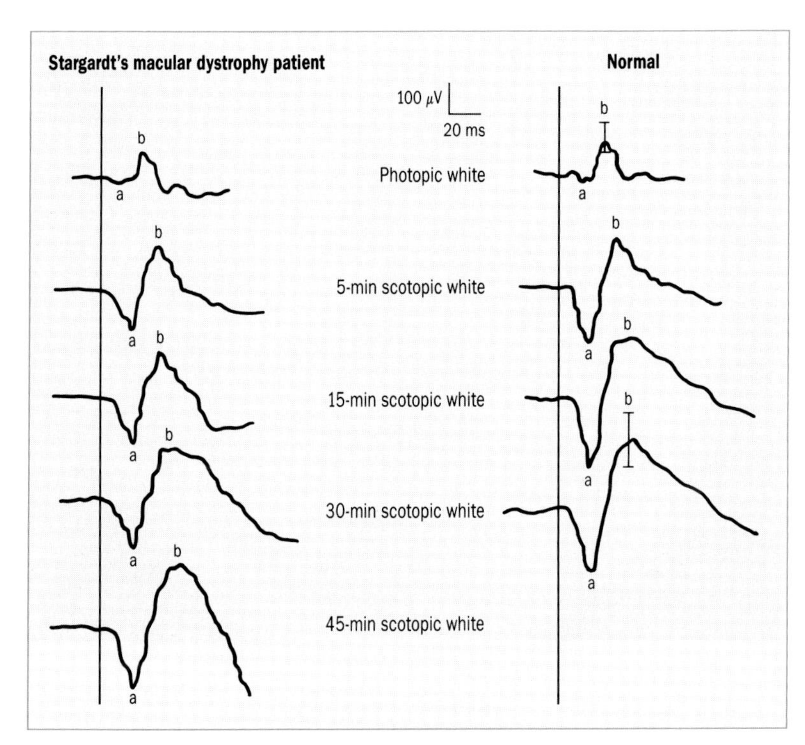

Figure 1-26 *ERG recording from a patient with Stargardt's macular dystrophy. Note that scotopic amplitude responses may not reach normal values unless ample time is allotted for dark adaptation. This period can exceed the usually sufficient 30 minutes.*

genetic disorders is probably spurious.

One study has classified patients with fundus flavimaculatus or Stargardt's macular dystrophy into four stages based on the extent of fundus changes and electrophysiologic findings.[198] In general, reduced ERG cone and rod a- and b-wave amplitudes in patients with Stargardt's macular dystrophy can be predicted by the extent of clinically apparent fundus pigmentary changes. Thus, patients with a localized atrophic-appearing foveal lesion with a ring or often a garland of more extensive fundus flecks show normal cone and rod ERG a- and b-wave amplitudes. Once numerous fundus flecks resorb and more diffuse retinal pigment epithelial atrophic changes occur, these amplitudes become reduced. It is imperative, however, to allow at least 45 minutes for dark adaptation to appreciate fully the optimal potential for rod ERG amplitudes, since patients with Stargardt's macular dystrophy can take longer than normal to obtain maximally dark-adapted amplitudes (Figure 1-26). Infrequently, some patients with a lesion clinically localized within the macula show a reduction in cone ERG a- and b-wave amplitudes. It is likely that these patients are manifesting variable expressivity in the severity of the functional impairment of cones rather than a separate and distinct genetic subtype of the disease.[199,200]

A histopathologic report by Eagle et al[201] demonstrated the presence of a

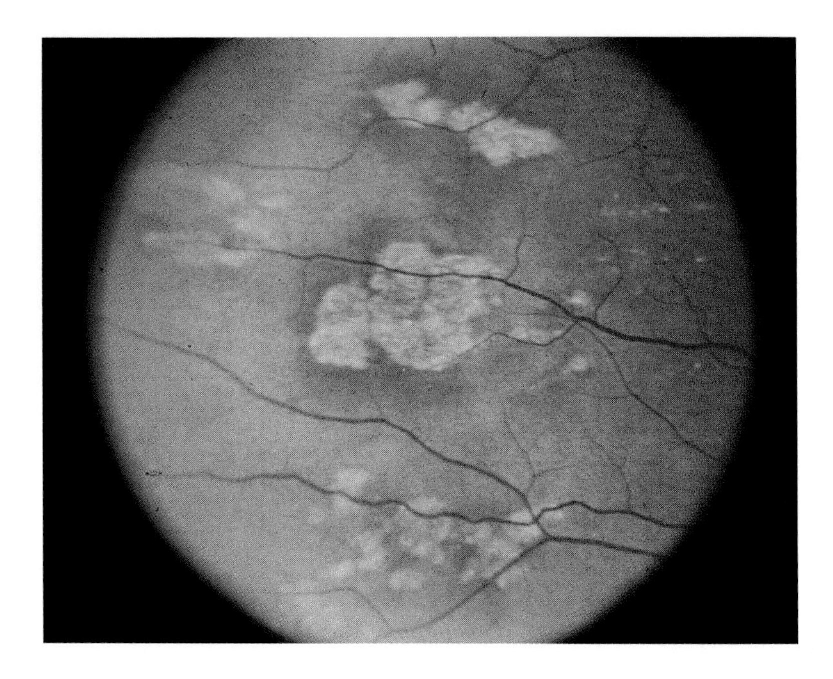

Figure 1-27 *Irregular fleck lesions characteristic in appearance to those seen in patients with fleck retina of Kandori.*
Courtesy of Donald Gagliano, MD.

lipofuscin-like material within all retinal pigment epithelial cells in a patient with fundus flavimaculatus. The fundus flecks corresponded to retinal pigment epithelial cells that had undergone hypertrophy as a consequence of particularly extensive accumulation of the lipofuscin-like material. The accumulation of this material within retinal pigment epithelial cells and the consequent effect on choroidal fluorescence may at least partially explain the dark choroid reported by Fish et al[202] in patients with Stargardt's macular dystrophy.

1-12-2 Fleck Retina of Kandori

This rare disorder, presumed to be autosomal recessively inherited, was first described in two patients by Kandori in 1959.[203] A generally mild to moderate degree of nonprogressive impairment of night vision is the presenting complaint. Central vision, color vision, and peripheral visual fields are normal.

The disorder is characterized by deeply located, irregular, sharply defined, "dirty"-appearing yellow flecks distributed primarily within the nasal equatorial region of the fundus and varying in size from the diameter of a small retinal blood vessel to 1.5 disc diameters (Figure 1-27). Macular lesions are not noted, and retinal vessels and the optic disc are normal. Clumping and migration of melanin pigment are not a feature of this disorder.

Electroretinographic examination with higher-intensity stimuli shows a negative pattern with a normal a-wave, but with the b-wave reduced in amplitude relative to the a-wave in scotopic as well as

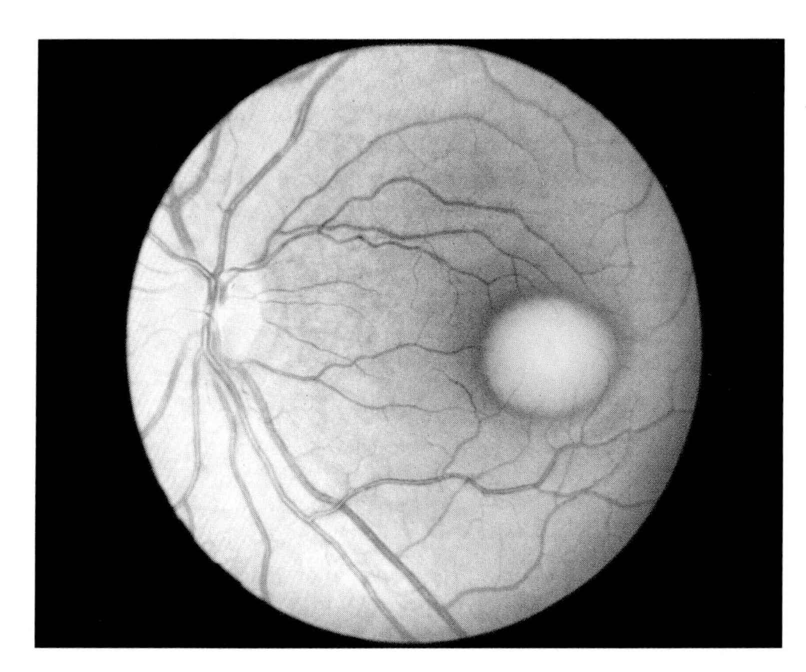

Figure 1-28 *Characteristic egg-yolk–like, sunny-side-up appearance of a macular lesion seen in the left eye of a patient with Best's vitelliform macular dystrophy.*

photopic conditions.[204] No time delay in recovery of the dark-adapted a- or b-wave amplitude is seen, although dark-adaptation testing shows a slight delay in reaching ultimately normal dark-adapted final thresholds. Electro-oculogram ratios are reported as normal.

The fundus lesions would seem to implicate a disorder at the level of the retinal pigment epithelium, while the negative ERG amplitude profile suggests functional impairment within the more proximal retina.[205]

1-12-3 Best's (Vitelliform) Macular Dystrophy

This autosomal dominantly inherited macular disorder, first described by Friedrich Best in 1905, presents with a pleomorphic ophthalmoscopic picture. Typical initial macular lesions, which are most frequently diagnosed between the ages of 5 and 15 years, demonstrate a yellow egg-yolk–like, "sunny-side-up" appearance (Figure 1-28). The presence of this macular lesion is still consistent with visual acuity most frequently 20/25 or better. Although only a single isolated foveal lesion is usually present, multiple nonfoveal foci of yellow egg-yolk–like lesions can be seen. Progressive impairment of visual acuity tends to parallel a subsequent course in which the yellow material appears to "rupture" or become fragmented into a "scrambled-egg" appearance. Eventually, the scrambled-egg appearance is replaced by a fibrotic (gliotic), hypertrophic-appearing scar. In some patients, the yellow material may resorb and subsequently be resecreted. In other patients, subretinal neovascularization may infrequently develop. Most patients are visually asymptomatic or show only slight to moderate visual loss until between 40 and 50 years of age. Peripheral vision remains normal. Electroretinographic cone and rod a- and b-wave amplitudes are typically normal in patients with Best's vitelliform macular dystrophy. However, Nilsson and Skoog[206] as well as Rover et al[207] reported either absent or small ERG c-wave amplitudes in patients with this disorder.

The definitive diagnostic test in this disease is the electro-oculogram (EOG), which is markedly abnormal in affected patients. Deutman[208] noted abnormal EOG light-peak to dark-trough ratios even when lesions were not ophthalmoscopically apparent. Weingeist et al[209] observed an abnormal accumulation of a lipofuscin-like material within all retinal pigment epithelial cells of a patient with Best's macular dystrophy, a finding also noted by O'Gorman et al[210] in another such patient.

An ERG will not be of any diagnostic value in a patient with a characteristic egg-yolk–appearing lesion. On occasion, a patient with Best's macular dystrophy can have bilateral atrophic-appearing foveal lesions that may show some phenotypic similarities to other hereditary macular lesions, such as those seen in patients with Stargardt's macular dystrophy or cone dystrophy. However, the absence of fundus flecks and a dark choroid, seen in Stargardt's macular dystrophy, as well as a markedly abnormal EOG helps distinguish patients with Best's macular dystrophy from those with Stargardt's macular dystrophy. The normal ERG and abnormal EOG in patients with Best's macular dystrophy differentiate these patients with an atrophic foveal lesion from those with cone dystrophy, who, with few exceptions, manifest abnormal cone ERG a- and b-wave amplitudes but a normal EOG response.[196,200]

1-12-4 Butterfly Dystrophy

Patients with butterfly dystrophy show a pigmentation or accumulation of a yellow-white substance within the macula that morphologically resembles a butterfly (Figure 1-29).[211] Some patients have been reported with abnormal EOG ratios, although this finding has not been consistent. The ERG has normal photopic and scotopic functions. As in Best's dystrophy, the lesions are bilateral and generally symmetric. This disorder has been classified as a pattern dystrophy.[212] Patients tend to maintain good or reasonably good central vision. In isolated instances, marked atrophy of the retinal pigment epithelium and choriocapillaris within the macula can be associated with appreciable impairment in central acuity.[213]

1-13

CHORIORETINAL DYSTROPHIES

1-13-1 Choroidal (Choriocapillaris) Atrophy

Cases of choroidal (choriocapillaris) atrophy can occur in the following forms: (1) central areolar, (2) central, (3) peripapillary, and (4) diffuse. All can be inherited as either autosomal dominant or autosomal

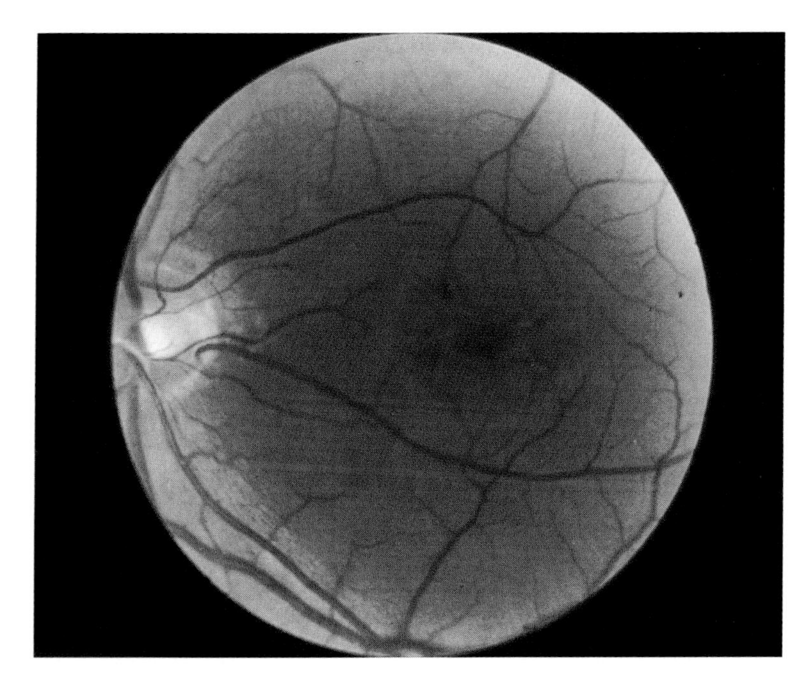

Figure 1-29 *Pigmented, butterfly-shaped foveal lesion in the left fovea of a patient with butterfly dystrophy. Courtesy of August F. Deutman, MD.*

recessive traits. The latter three categories may represent a continuum of the same disease and not separate genetic diseases, since their expression may be interrelated in some families. Most frequently, after approximately age 40–50 these patients show a decrease in acuity and, in the diffuse form, poor night vision. The ERG is subnormal in the peripapillary and diffuse forms, becoming nondetectable with more advanced disease. Generally, the amplitude of the ERG parallels the clinically apparent fundus involvement. Occasionally, however, cases of central and central areolar choroidal atrophy show lower amplitudes than might be anticipated from the extent of the manifest fundus pathology. In these instances, consideration should be given to the diagnosis of cone dystrophy. In contrast to patients with retinitis pigmentosa, the implicit times are generally normal in patients with choroidal atrophy.

1-13-2 Gyrate Atrophy of the Choroid and Retina

Patients with gyrate atrophy, an autosomal recessive chorioretinal dystrophy, generally present between 20 and 30 years of age complaining of poor night vision. The primarily peripheral and midperipheral fundus has multiple, initially discrete, irregular, atrophic-appearing patches of pigment epithelium, choriocapillaris, and, later, larger choroidal vessels (Figure 1-30). The lesions tend to become confluent as they extend both centrally

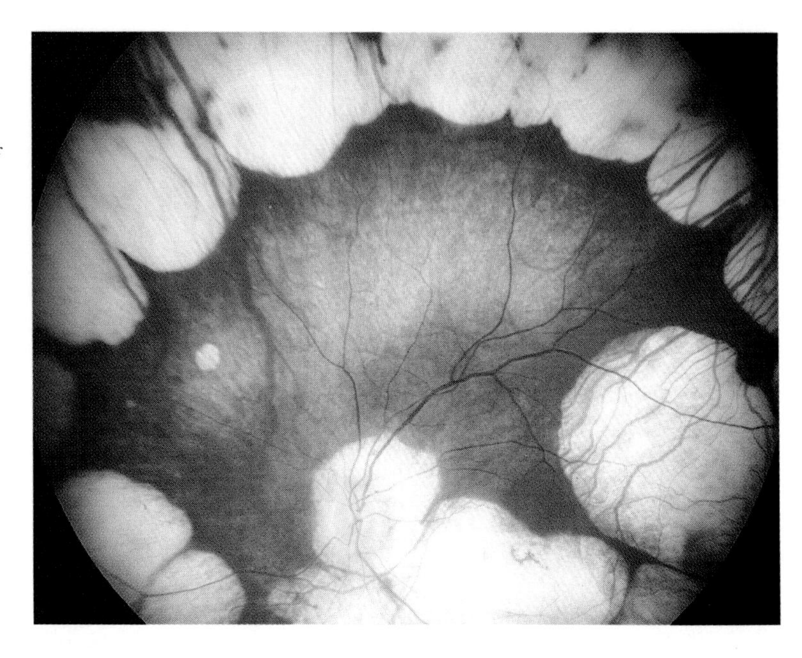

Figure 1-30 *Characteristic atrophic-appearing areas of the retinal pigment epithelium and choroid in the left eye of a patient with gyrate atrophy of the choroid and retina.*

and peripherally. Patients are generally myopic and frequently have lens opacities. The ERG cone and rod amplitudes are usually either markedly reduced or nondetectable. In addition to night blindness, patients experience progressive impairment in peripheral field and central vision. Biochemical abnormalities include 10- to 20-fold elevations in plasma ornithine, hypolysinemia, hyperornithinuria, and either an absence or a marked reduction in ornithine alpha-aminotransferase (OAT) activity in cultured skin fibroblasts and in lymphocytes.[214] Some patients when administered pyridoxal phosphate (vitamin B_6) orally show a 20–50% reduction in plasma ornithine levels, while others are nonresponders to B_6.[215] Gyrate atrophy of the choroid and retina is one of the few progressive night-blinding disorders in which a metabolic defect has been implicated and for which therapeutic trials with a low protein diet are under investigation.[216]

1-13-3 Choroideremia

The X-linked recessive chorioretinal dystrophy choroideremia was first described in 1871 by Mauthner.[217] Affected males complain of poor night vision within the first 10–20 years of life and show bilateral progressive degeneration of the choroid and retina. Although extensive degenerative changes of the retinal pigment epithelium as well as the choroid become apparent, unlike the pigmentary changes of the retina seen in retinitis pigmentosa patients, there is little or no migration of

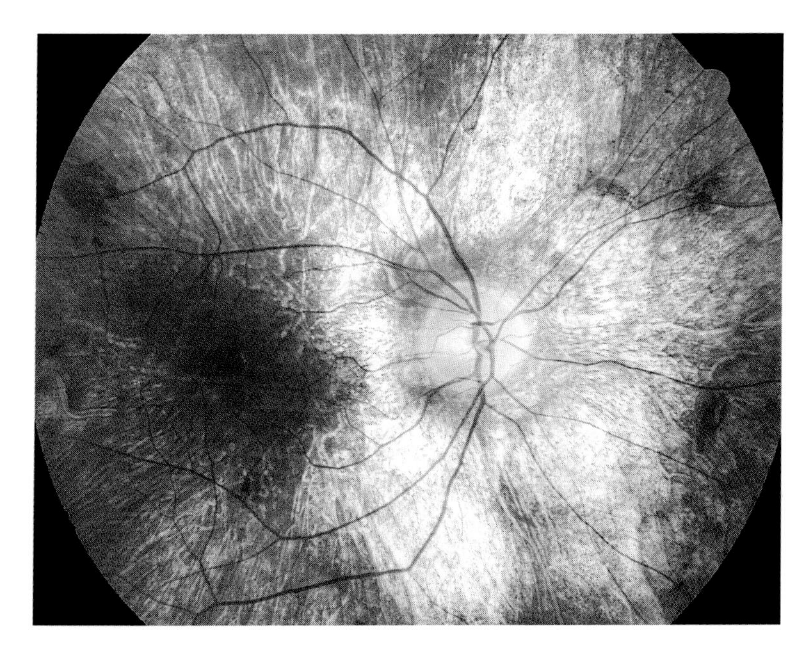

Figure 1-31 *Extensive atrophy of the retinal pigment epithelium and choroid in a patient with choroideremia. Eventually, even more extensive atrophy of the choroid occurs in such patients.*

bone-spicule–like pigment into the anterior layers of the retina (Figure 1-31).

Initially, central acuity is normal or nearly so, but it eventually undergoes a progressive deterioration. Peripheral fields are usually moderately depressed even at an early age, with severe peripheral field restriction occurring in the 5th and 6th decades of life. Even in the initial stages of the disease, the ERG and EOG are markedly abnormal. When ERG responses are recordable, they tend to be primarily residual cone function.

The fundus of carrier females may show a "moth-eaten" appearance of the retinal pigment epithelium in the presence of normal photoreceptor cell function as determined by ERG recordings.[218]

Genetic linkage studies have demonstrated that a DNA fragment polymorphism (DXYS1), located on the long arm of the X chromosome at the locus Xq 13-21, shows reasonably close linkage with the X-linked gene for choroideremia.[219,220]

1-14

BRUCH'S MEMBRANE DISORDER (ANGIOID STREAKS)

Angioid streaks have been associated most frequently with pseudoxanthoma elasticum, Paget's disease, and sickle cell disease. The ERG can be subnormal in more advanced cases. An abnormal EOG has also been noted in those cases with more advanced fundus changes. François and de Rouck[221] reported subnormal ERG responses in 14 (46%) of patients with angioid streaks found in association with pseudoxanthoma elasticum (Groenblad-Strandberg syndrome). In 8 cases, both cone and rod systems were affected, in 4 only the rods were affected, and in 2 only the cones were affected. These changes were found most often in eyes with advanced disciform macular degeneration. Normal ERG records, however, were obtained even in the presence of advanced macular degeneration.

1-15

HEREDITARY VITREORETINAL DISORDERS

The three most frequently seen disorders in this category include (1) X-linked juvenile retinoschisis, (2) Favre-Goldmann syndrome, and (3) Wagner-Stickler degeneration. All of these entities have both vitreous and retinal abnormalities. Patients with X-linked juvenile retinoschisis have peripheral retinoschisis in approximately 40–50% of cases, foveal microcystic changes in virtually all cases (Figure 1-32), and, less frequently, atrophic-appearing macular lesions. The vitreous either is optically empty or contains fibrous bands. The ERG has a reduced photopic and scotopic b-wave amplitude, reflecting functional impairment within the inner retinal layers (Figure 1-33). The a-wave, although not infrequently subnormal in amplitude,[222] is considerably less impaired than the b-wave, creating a negative ERG pattern. Table 1-2 lists other entities that can selectively or predominantly affect the ERG b-wave amplitude while maintaining a normal or near-normal a-wave amplitude. Oscillatory potentials are markedly reduced in amplitude or are not detectable in retinoschisis patients, while EOG light-peak to dark-trough ratios are normal in approximately 90%.[222] In later stages, patients with X-linked juvenile retinoschisis can manifest a diffuse, nonspecific retinal pigmentary degeneration with areas of pigment atrophy and clumping.[223] At this later stage, marked impairment of both a- and b-wave ERG amplitudes, as well as EOG ratios, is found.[223]

Lewis et al[224] described 3 female patients with a familial foveal retinoschisis showing phenotypic similarity to the foveal lesions seen in X-linked juvenile retinoschisis. The essentially normal or only minimally subnormal ERG, particularly the absence of a selective or predominant b-wave amplitude reduction, was distinctly different from the negative-type

TABLE 1-2

Entities With Selective or Predominant Decrease in b-Wave Amplitude

Siderosis
Central retinal artery or vein occlusion
X-linked juvenile retinoschisis
Congenital stationary night blindness
Oguchi's disease
Myotonic dystrophy
Lipopigment storage disorders (neuronal ceroid lipofuscinosis, Batten's disease)
Coats' disease
Quinine intoxication
Methanol intoxication

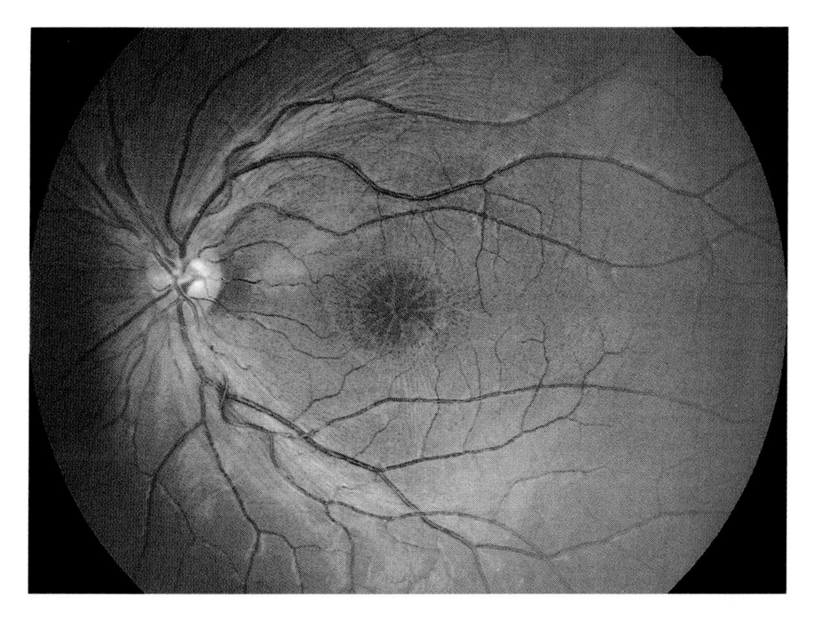

Figure 1-32 *Foveal microcystic changes as characteristically seen in a patient with X-linked juvenile retinoschisis.*

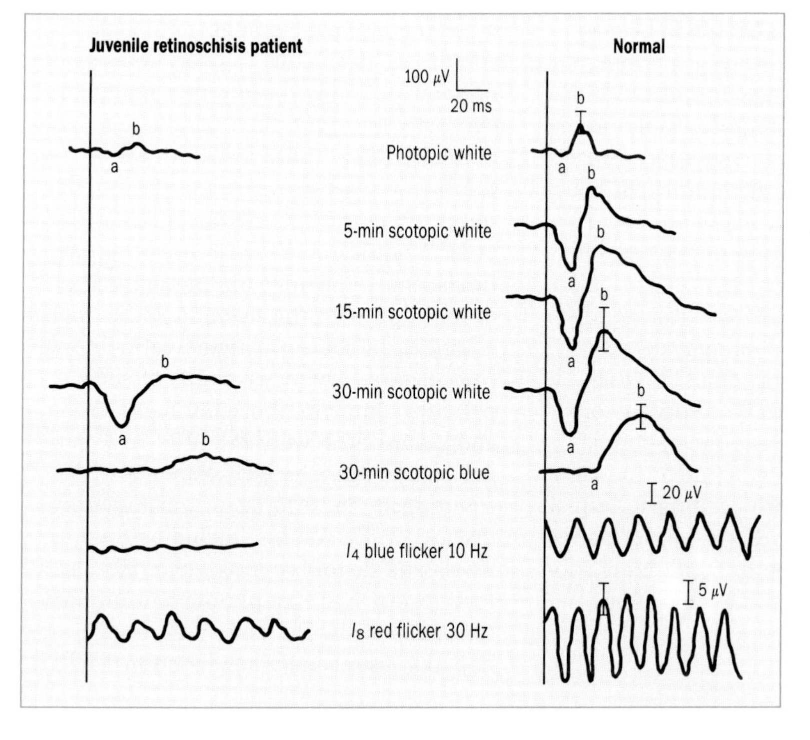

Figure 1-33 *ERG recording from a patient with X-linked juvenile retinoschisis, showing selective b-wave amplitude reduction to high-luminance white light flash stimulus.*

ERG pattern seen in patients with X-linked juvenile retinoschisis. However, Shimazaki and Matsuhashi[225] reported a negative-type ERG in a mother and daughter, both of whom showed foveal cystic changes and peripheral retinoschisis. They were believed to have a dominantly inherited disorder distinct from X-linked juvenile retinoschisis. Peripheral retinoschisis without foveal schisis was noted in a father and daughter in whom the ERG showed a small reduction in both a- and b-wave amplitudes without a predominant reduction in the b-wave.[226] Noble et al[227] reported foveal retinoschisis and poor night vision in 2 patients who showed a generalized rod-cone dystrophy by ERG recordings.

The Favre-Goldmann syndrome is an autosomal recessive disease in which patients characteristically complain of poor night vision and show atypical peripheral pigmentary degeneration, macular cystic degeneration, peripheral (often inferotemporal) retinoschisis, posterior subcapsular lens opacities, an optically empty vitreous with an occasional vitreous band, and a markedly abnormal, often nondetectable ERG.[228]

Wagner-Stickler vitreoretinal degeneration is an autosomal dominant disease characterized by lens opacities, a vitreous that either is optically empty or contains vitreous bands, pigmentary retinal changes, peripheral degenerative retinal changes not infrequently leading to retinal detachment, and an ERG response that is either normal or mildly to moderately subnormal, varying with the extent of apparent fundus involvement.[229] Patients are frequently myopic. In some families, members show facial and other skeletal anomalies traditionally described in association with Stickler's syndrome. Patients who present with an autosomal dominant disorder and the ocular findings described should be comprehensively examined for the presence of any orofacial or musculoskeletal abnormalities.

1-16

INFLAMMATORY CONDITIONS

The degree to which various inflammatory conditions of the choroid and retina affect the ERG most often depends directly on the extent of apparent fundus abnormality. Thus, patients with minimal to moderate fundus involvement from such conditions as syphilis or chorioretinitis of unknown etiology have either normal or generally not more than moderately subnormal ERG amplitudes. Both the a- and the b-waves are affected and the implicit times are normal.[230] However, Cantrill et al[231] noted prolonged b-wave implicit times, most apparent under photopic conditions and with the use of a 30-Hz flicker stimulus, in certain patients with chronic pars planitis. Over two thirds of such patients complained of poor night vision, and almost 50% were found to have retinal pigmentary changes. Patients with traditionally local forms of inflammatory disease of the fundus, including toxoplasmosis and histoplasmosis, generally manifest normal ERG amplitude and implicit time responses. Patients with rubella retinopathy also usually have a normal ERG,

A

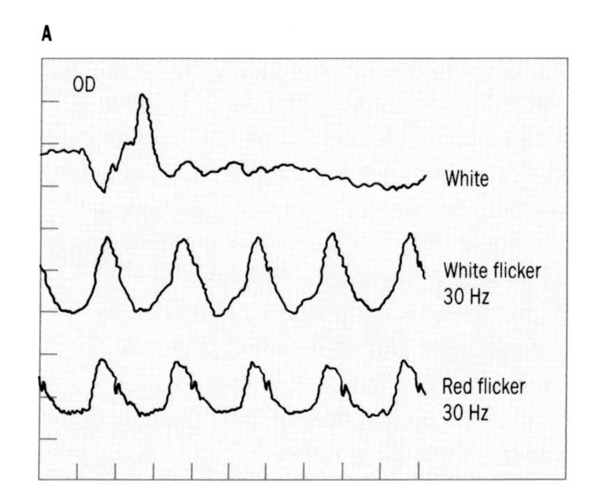

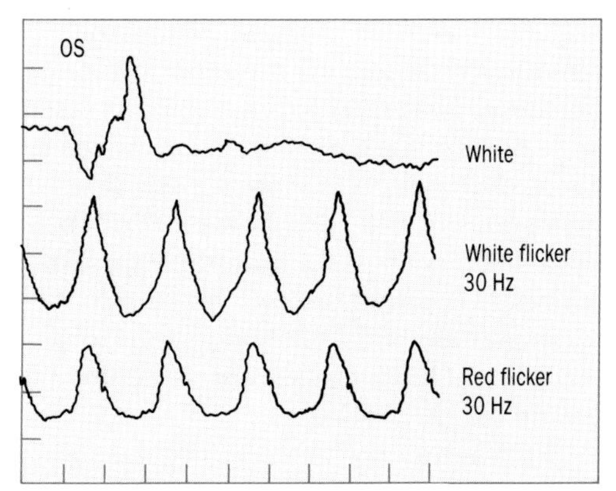

B

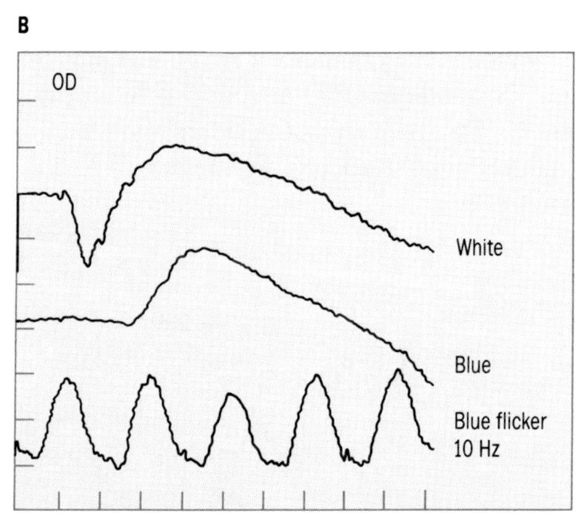

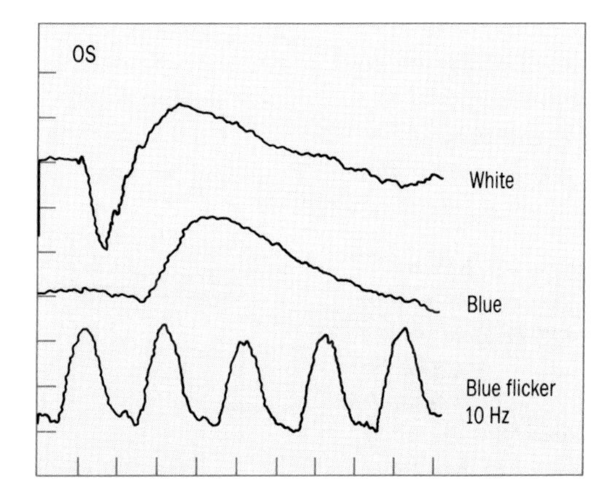

since the pathology appears to involve the melanin granules within the pigment epithelial layer of the retina. However, when interocular asymmetric pigmentary retinal changes of a patient are encountered, the eye with more apparent retinal changes may be found to show smaller ERG amplitudes than the less affected eye (Figure 1-34). The less extensive ERG abnormalities, generally found in inflammatory disorders of the choroid and retina that lead to pigmentary degenerative changes,

Figure 1-34 *ERG recording from a patient with asymmetric fundus pigmentary changes found in association with rubella. Pigmentary changes were more extensive in the right eye. Note that the patient's (A) light-adapted and (B) dark-adapted responses were of slightly less amplitude in the eye with more extensive pigmentary changes (OD).*

can be of value in evaluating these conditions, which, on occasion, mimic the fundus appearance of true retinitis pigmentosa. Nevertheless, in some presumed inflammatory disorders, functional impairment of the retina is more apparent, by reductions in ERG amplitudes, than might have been anticipated from the extent of clinically apparent disease. Two such examples are birdshot chorioretinopathy and the multiple evanescent white dot syndrome (MEWDS).

Birdshot retinochoroidopathy was initially described in 1980 and termed such because of unusual multiple cream-colored lesions, without hyperpigmentation, at the level of the choroid and retinal pigment epithelium of the primarily posterior pole of the fundus.[232] Vitritis, retinal vasculitis, and cystoid macular edema are also encountered. Other descriptive terms such as "vitiliginous chorioretinitis"[233] and "salmon-patch choroidopathy"[234] have been used. Reductions in ERG amplitudes have been reported in most but not all patients (Figure 1-35). A prolonged photopic and scotopic b-wave implicit time was reported by Fuerst et al,[235] although only amplitude reductions may be apparent during the initial stages of this disease.[236] In general, scotopic responses are more impaired than those obtained under photopic conditions, while a decrease in b-wave amplitude has been noted as greater than that for the a-wave.[235,237] Oscillatory potentials are either nondetectable or reduced in amplitude. The majority of patients show a reduction in the EOG light-peak to dark-trough ratio.[238]

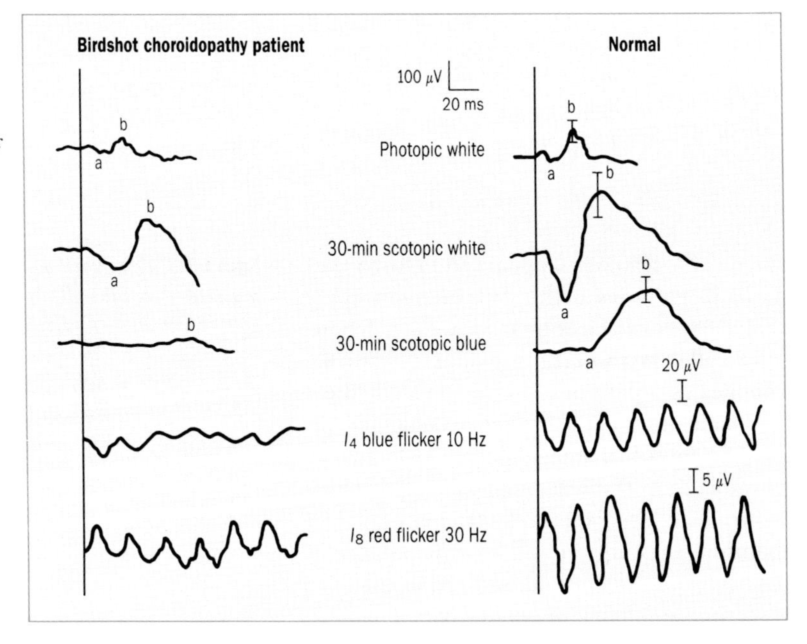

Figure 1-35 *ERG responses from a patient with birdshot choroidopathy. Note that the responses under scotopic conditions were more impaired than those obtained under photopic conditions.*

Patients affected with the multiple evanescent white dot syndrome present with an acute, generally unilateral onset of visual loss associated with the ophthalmoscopic appearance of small discrete white dots at the level of the retinal pigment epithelium or deep retina. A granularity of the fovea is characteristically found. Other clinical findings include the presence of vitreal cells, retinal vascular sheathing, and optic disc edema. These findings gradually regress over several weeks, and vision returns to normal or near-normal levels.[239] Electrophysiologic studies during the acute stage show marked reduction in ERG a- and b-wave amplitudes, followed by almost complete recovery. Early receptor potential recordings are similarly found to show an initial reduction in amplitude with subsequent recovery.[240]

1-17

CIRCULATORY DEFICIENCIES

1-17-1 Sickle Cell Retinopathy

In patients with sickle cell retinopathy, ERG a-wave, b-wave, and oscillatory potential amplitudes were found to be normal in the absence of peripheral retinal neovascularization. However, these ERG components were reduced in amplitude compared to normal when peripheral retina neovascularization was present.[241]

1-17-2 Takayasu's Disease

The chief characteristics of Takayasu's (pulseless) disease are (1) involvement usually of both eyes in females around 20 years of age; (2) arteriovenous anastomoses around the optic disc, with aneurysm-like vascular dilation and small hemorrhagic spots unassociated with any inflammatory conditions; (3) impaired vision, with later development of cataract; and (4) general circulatory disturbances, including undetectable radial pulses. The disease represents a primary aortitis, with spread to the surrounding vessels. In the early stages of pulseless disease, decreased or abolished oscillatory potentials without appreciable effect on the a- and b-wave amplitudes are seen. With progression, both a- and b-wave amplitudes can become reduced or even nondetectable.

1-17-3 Carotid Artery Occlusion

The effect of carotid artery occlusion on ERG a- and b-wave amplitudes depends on the extent and severity of the occlusion, its location, and the degree of collateral circulation that has developed with branches of the external carotid artery and through the circle of Willis via the anterior communicating artery. With occlusion of the internal carotid artery, a reduction in both a- and b-wave amplitudes can be found. The a-wave reduction is generally relatively less than that seen in the b-wave, which is more sensitive to ischemia. Krill and Diamond[242] found a consistently smaller b-wave associated with insufficient blood flow through the carotid artery. These changes are less likely to be seen if ample collateral flow through the circle of Willis and/or vertebrobasilar system has been established.

Collateral flow from anastomosis between branches of the external carotid artery and the ophthalmic artery can result in a reversed blood flow through the ophthalmic artery and a relative state of ischemia, which in turn causes a reduction in a- and b-wave amplitudes.[243] Occlusion or stenosis of the common carotid artery exerts a distinct reduction in ERG a- and b-wave amplitudes.[243] Johnson et al[244] emphasized delays in ERG a- and b-wave implicit times as well as in the 30-Hz flicker response in patients with venous stasis retinopathy found in association with carotid artery stenosis. These patients were also found to have a reduction in retinal sensitivity, as determined by elevations in the half-saturation constant of their intensity-response function (*see Section 1-5-2*).

1-17-4 Central Retinal Artery and Vein Occlusion

Retinal circulatory disturbances affect the ERG b-wave potential (Figure 1-36). Both a decrease in scotopic b-wave amplitude and diminished or nondetectable oscillatory potentials accompany central retinal artery occlusion. Henkes[245] noted this b-wave reduction in 21 of 24 cases, and Karpe and Uchermann[246] reported similar findings in 15 of 16 cases. Flower et al[247] reported an increased susceptibility to

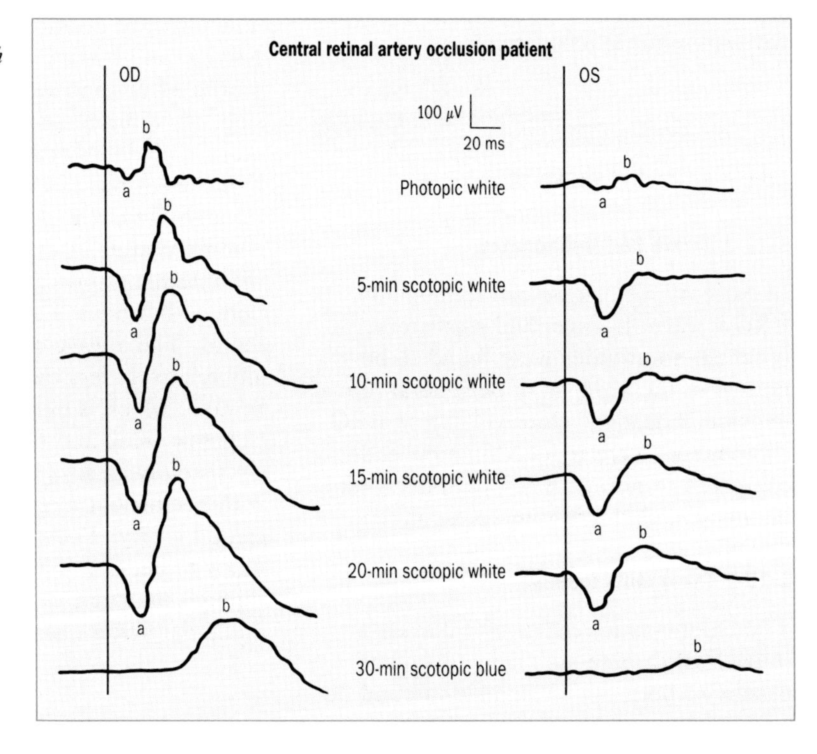

Figure 1-36 *ERG recording from a patient with central retinal artery occlusion. Note the selective reduction of b-wave amplitude in the affected left eye.*

hypoxia of the light-adapted compared to the dark-adapted b-wave in a patient with central retinal artery occlusion. Central retinal vein occlusions can cause a similar reduction in the ERG b-wave amplitude. Karpe and Uchermann[246] studied 73 cases of central vein occlusion and found a reduced scotopic b-wave amplitude in 44 cases and a normal or increased scotopic b-wave amplitude in 29 cases. A reduction of the scotopic b-wave amplitude in patients with central retinal vein occlusion was also noted by Eichler and Stave.[248] In branch artery occlusions, there is generally either a slight b-wave reduction or a normal ERG response. Ponte[249] reported the ERG normal in 52%, subnormal in 38%, subnormal negative in 8%, and supernormal in 2% of patients with branch artery occlusion. Ponte noted no alteration in the ERG in 61% of patients with branch vein occlusions, while 32% had a subnormal negative pattern; the finding of a supernormal response was exceptional, seen in only 1% of such cases. In branch vein occlusion, the visual prognosis correlated directly with the effect on oscillatory potential amplitudes. In central retinal vein occlusion, a favorable prognosis correlated more closely with a supernormal or normal ERG than either a subnormal or a negative ERG, which was presumably associated with a greater degree of hypoxia. Karpe and Uchermann[246] found no definite correlation between the initial size of the b-wave amplitude and the subsequent visual acuity after photocoagulation therapy in patients with retinal vein occlusion.

Johnson et al[250] compared eyes having central retinal vein occlusion that developed iris neovascularization with eyes that had not developed neovascularization. Those developing iris neovascularization had a significantly reduced ERG amplitude and an increase in the stimulus luminance required to elicit a response equal to one half the maximal potential response (σ). All eyes with neovascularization showed large a- and b-wave and 30-Hz implicit time delays. These authors concluded that the delays in implicit times and amplitude reductions were more sensitive in ascertaining the presence of iris neovascularization than was either the ERG b- to a-wave amplitude ratio, as advocated by other investigators,[251] or fundus fluorescein angiography. Breton et al[252] emphasized that ERG amplitude and b- to a-wave ratios, as well as half-saturation intensity (log K or σ) and 30-Hz implicit time delays, were of value as predictors of rubeosis in central retinal vein occlusion patients. These authors noted that a multiple discriminant analysis, combining information from several ERG parameters, facilitated separation of central retinal vein occlusion patients in whom rubeosis would develop from those in whom it would not.

1-17-5 Hypertension and Arteriosclerosis

In patients with systemic hypertension and a normal fundus, Henkes and van der Kam[253] found supernormal responses in 9 of 15 patients. In patients with systemic hypertension and mild hypertensive retinopathy, ERG amplitudes can be either supernormal or normal. Reduction in oscillatory potential amplitudes may occur prior to, or be more apparent than, reductions of a- or b-wave amplitudes.[254,255] Those patients with evidence of more severe fundus hypertensive retinopathy frequently have either subnormal a- and b-wave responses or subnormal b-wave amplitudes. Henkes and van der Kam[253] further noted that ERG b-wave amplitude reduction occurs in patients with retinal arteriosclerotic changes without associated arterial hypertension. Occasionally, both a- and b-waves are subnormal. The b-wave amplitude reduction secondary to hypoxic retinal changes is generally appreciated more under dark-adapted than under light-adapted conditions.

1-18

TOXIC CONDITIONS

1-18-1 Chloroquine and Hydroxychloroquine

The ocular side effects associated with prolonged use and/or high doses of either chloroquine diphosphate (*Aralen*, *Resochin*), chloroquine sulfate (*Nivaguine*), or hydroxychloroquine (*Plaquenil*) include impaired central vision, peripheral field depression, attenuated retinal vessels, both atrophic and granular pigmentary changes of the macula with the formation of a characteristic bull's-eye appearance, and, in some cases, eventual peripheral pigmentary changes and optic atrophy. Visual loss can progress even after the discontinuation of these antimalarial drugs. Verticillate, whorl-like changes within the corneal epithelium are not infrequently encountered. The reported incidence of chloroquine-induced retinal toxicity has varied, depending on the definition of retinopathy and the methods used for its detection. The incidence is probably from less than 1% to 6%,[256,257] particularly if adherence to currently recommended maximal daily dosages of 250 mg of chloroquine or 400 mg of hydroxychloroquine are maintained. The development of macular changes has been reported to occur more frequently with the use of chloroquine than with hydroxychloroquine. This finding may partially reflect the higher dosages of chloroquine that were used when this drug was initially introduced. Nevertheless, macular lesions can also develop in a small percentage of patients who use hydroxychloroquine.

In the stage of chloroquine retinopathy when degenerative changes are clinically apparent only in the macula, the ERG is usually normal or occasionally subnormal to a small degree. With more advanced and extensive disease, when peripheral pigmentary changes also become apparent, the ERG is usually moderately subnormal, while nondetectable or minimal responses are obtained in patients with far advanced retinopathy. The cone ERG function initially tends to be affected more than the rod function. The ERG is not an effective test to determine a minimal degree of functional impairment of the retina in cases of early chloroquine toxicity.[258-260] Static perimetry testing within the central 10° is more likely to ascertain early functional impairment than is a test of diffuse retinal cell dysfunction such as the ERG.

1-18-2 Chlorpromazine

Siddall[261] described a dusky, granular appearance to the fundus from fine pigment clumping in patients receiving chlorpromazine (*Thorazine*). A few patients had a more coarse pigment clumping, while one had actual pigment migration. The toxic level of chlorpromazine causing abnormal retinal pigmentation was estimated to be 2400 mg per day taken for 8–12 months. The pigment accumulation became less apparent or resolved as the dosage was reduced or the drug discontinued. Retinal pigmentary changes from the use of chlorpromazine are significantly less severe than are the coarse plaques of pigment deposition seen in thioridazine (*Mellaril*) chorioretinopathy.

Chlorpromazine probably has significantly less toxic effects on the retina than either thioridazine or chloroquine. Alkemade[262] found that 15 of 99 patients given chlorpromazine developed a retinopathy. The incidence of the retinopathy increased with greater daily dosages and longer duration of treatment. The drug has a selectivity for melanin-containing cells of the choroid particularly and is only slowly eliminated from the body. Boet[263] cited Rintelen's findings of no abnormalities in either the fundus, on dark adaptation, or the visual fields in patients receiving chlorpromazine. Boet[263] also discussed the findings of Baumann on 35 schizophrenic patients treated with chlorpromazine. No abnormalities of either dark adaptation, visual fields, color vision, or fundus examination were noted in 24 cases. Of the remaining 11 cases, 9 had slight abnormalities in visual fields and dark adaptation, which were attributed to opacities of the media or the patient's mental condition. Fundus pathology, however, was not present. The other 2 patients apparently had a slightly exaggerated pigment mottling. Busch et al[264] in 1963 found no ocular abnormalities in 363 patients treated with chlorpromazine and promethazine hydrochloride. Only visual acuities and results of the fundus examination were recorded. Although final conclusions of the effects of chlorpromazine on the ERG await further extensive reports, it appears that toxic effects on the photoreceptors are not an early consequence of treatment with chlorpromazine. It might be anticipated that, in later stages, associated with coarse pigment clumping and migration, the ERG would show reductions in amplitude, as noted in thioridazine retinal toxicity.

1-18-3 Thioridazine

Thioridazine hydrochloride (*Mellaril*) was introduced in 1959 into clinical trials for the treatment of psychoses. Patients receiving relatively high doses of this drug can experience decreased visual acuity and night blindness. Both central and ring scotomas have been noted. In earlier stages, a pigmentary mottling appears in the macular and perimacular regions. Later, extensive degenerative changes of the pigment epithelium, photoreceptors, and choriocapillaris are seen (Figure 1-37).[265] The ERG shows various degrees of diminished photopic and scotopic a- and b-wave responses that correlate with apparent fundus changes (Figure 1-38). Potts[266] reported that the uveal pigment is the ocular tissue that most highly

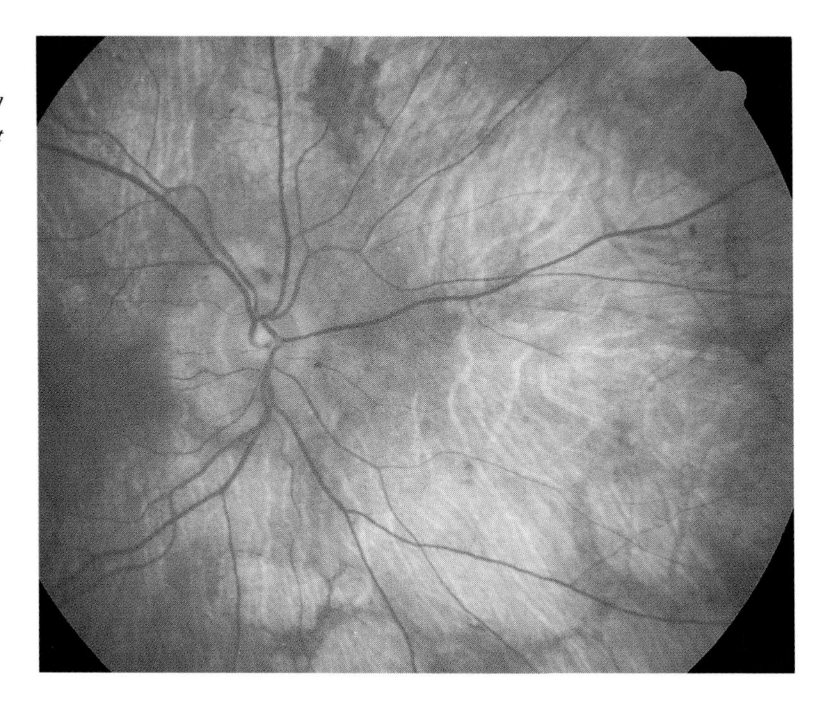

Figure 1-37 *Extensive atrophy of the retinal pigment epithelium and the choroid in a patient with Mellaril-induced retinal toxicity. Nummular, or coin-shaped, lesions are characteristic, at least at some stage.*

concentrates this phenothiazine derivative. As a consequence, the choriocapillaris, retinal pigment epithelium, and photoreceptors undergo degenerative changes that can progress for several years, even after the discontinuation of therapy.[267,268] In contrast, Appelbaum[269] studied 77 patients receiving thioridazine and found no pigmentary retinopathy suggestive of drug toxicity. ERG studies, however, were not obtained. While opinions vary, 1600 mg of thioridazine per day is probably a safe therapeutic dosage and 800 mg per day, over approximately 20 months, is probably a maximally safe maintenance dosage to avoid a degenerative chorioretinopathy. A phenothiazine preparation closely allied to thioridazine is NP-207. This experimental drug, which is the chlorine analogue of thioridazine, was found to produce night blindness, loss of visual acuity, and a severe pigmentary retinopathy with reductions in ERG amplitude.[270]

1-18-4 Indomethacin

Henkes et al[271] indicated that even with prolonged administration of indomethacin (*Indocin*), ophthalmic side effects rarely occur. They did report, however, the case of a 38-year-old man who, during the use of indomethacin, noted a deterioration of visual acuity, visual fields, and dark adaptation and showed a reduced scotopic ERG b-wave amplitude and EOG light-to-dark ratio. Pigmentation was scattered throughout the retina. After discontinua-

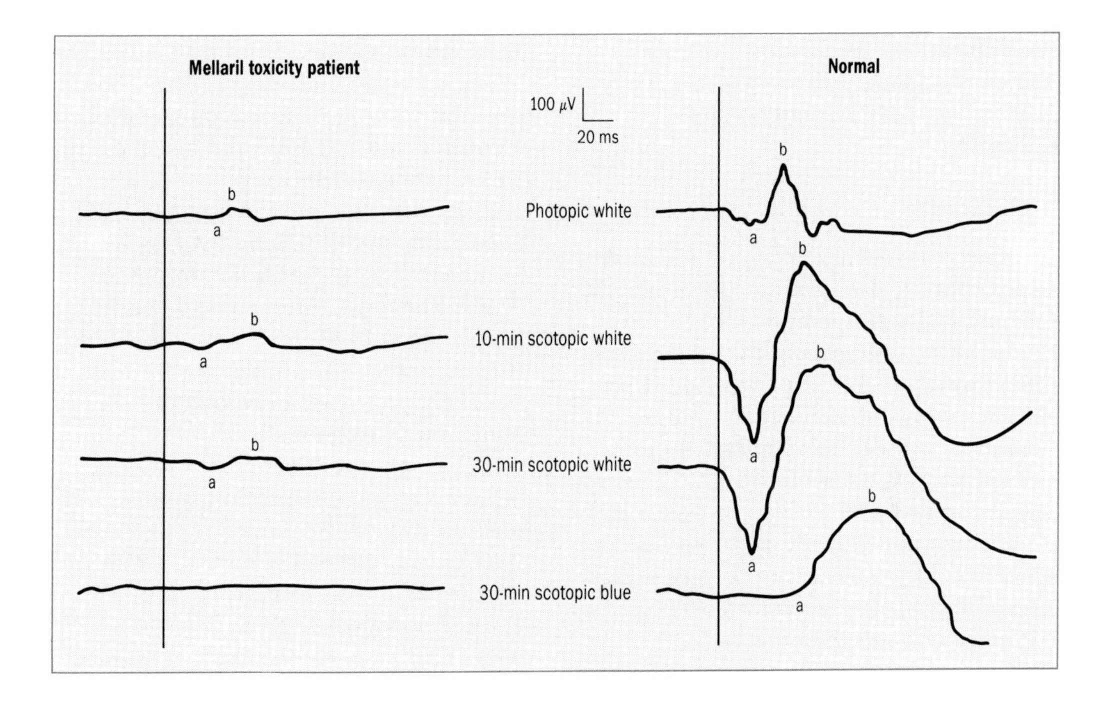

Figure 1-38 *Marked reductions in ERG amplitudes noted in a patient with Mellaril-induced retinal toxicity. ERG recordings can be used to monitor progressive impairment in retinal function that can occur even after the drug is discontinued.*

tion of therapy, these visual functions improved over a period of several months. Rothermich[272] reported an ophthalmologic study of 263 arthritic patients, 111 of whom received indomethacin. No difference in the incidence of ocular changes was noted between those patients receiving indomethacin and those not receiving the drug. The longest time any patient received the medication was 5 years. Burns[273] reported ophthalmic side effects of indomethacin that, in addition to subepithelial corneal deposits, included an abnormal fundus appearance in 15 of 34 patients. Moreover, 6 patients showed paramacular depigmentation varying from a mottled appearance to involvement with extensive areas of pigment atrophy. Burns further found a disturbance of certain visual functions, as demonstrated by abnormal scotopic ERG a- and b-wave amplitudes, dark adaptation, and perimetry.

1-18-5 Quinine

Quinine is a cinchona alkaloid used in the treatment of malaria and nocturnal muscle cramps. Acute poisoning induces a syndrome known as cinchonism, usually seen after attempted abortion or suicide. Findings in this syndrome include tinnitus, headache, nausea, gastrointestinal upset, tremor, and hypotension. Adverse effects usually occur with ingestion in excess of 4 g, although idiosyncratic sensitivity is documented. Death has occurred with ingestion of 8 g. Visual symptoms usually occur within 2–24 hours after ingestion.

The fundus changes in the initial stages of quinine toxicity are characterized by retinal edema, causing a grayish white fundus appearance with a cherry-red macula. Visual acuity is decreased and peripheral fields are restricted. Pupillary dilation is a consistent feature of acute poisoning. Although a transient subnormal ERG response of both a- and b-wave amplitudes is apparent if testing is done within the first 12–18 hours, the ERG is most frequently normal when recordings are obtained after 24 hours. Initial reports attributed the pathology of the early stages to involvement of the ganglion cells, but subsequent findings by Cibis et al[274] in rabbits suggest that there is an early effect on the ERG from involvement of the outer layers of the retina. In the later stages as seen in human patients, visual acuity improves, as does the peripheral field impairment. At this stage, however, the arterioles become progressively attenuated and the optic disc becomes pale. Commensurate with these latter changes, the photopic and scotopic ERG b-wave amplitudes become progressively decreased. The flicker fusion responses are also abnormal. After approximately 1 month, the decrease in b-wave amplitude becomes stable. With time, the b-wave may slightly increase in amplitude, although rarely to normal values. Moloney et al[275] emphasized that cone ERG function remains more impaired than does rod function. Generally, the a-wave and the EOG remain essentially normal, although François et al[276] as well as Bacon et al[277] reported abnormal EOG findings in 3 cases and 1 case, respectively, of quinine intoxication. Moloney et al[275] reported a patient with marked EOG impairment initially who demonstrated slow recovery of function over a 2-year period.

1-18-6 Methanol

In acute methanol poisoning, Karpe[3] noted an increased a-wave and a reduced b-wave. Potts et al[278] confirmed this observation in rhesus monkeys. However, Ruedemann[279] found a distinct reduction in both a- and b-wave amplitudes, as well as increased implicit times. In these patients in the early stages, the fundus shows visible retinal and optic disc edema with marked visual reduction. In late stages, there is visible optic atrophy with restricted peripheral fields. In these later stages, both a- and b-waves are frequently subnormal. The retinal pathology is not well defined, especially with reference to the photoreceptors. Damage to the ganglion cells of the brain and retina has been demonstrated. The ERG findings also suggest a malfunction of the outer retinal layers. Methyl alcohol ocular toxicity has been assessed in the rhesus monkey, and the optic disc edema was postulated as due to axoplasmic stasis.[280,281]

1-18-7 Gentamicin

The retinal toxicity of intravitreally injected gentamicin (*Gentamycin*) in the rabbit eye has been assessed by several investigators. Marked reductions in ERG amplitudes can be found, although the dosage at which retinal toxicity occurs has differed between reports.[282-284] Palimeris et al[283] found an intravitreal injection of as little as 0.1 mg to be toxic to the rabbit retina.

1-19

VITAMIN A DEFICIENCY

Dhanda[285] studied scotopic responses in patients with vitamin A deficiency. Subnormal to nondetectable responses were noted in patients with xerosis and night blindness. After vitamin A administration, symptoms resolved and ERG values returned to normal. In general, children below the age of 15 years are particularly prone to develop night blindness as a result of vitamin A deficiency. Contrary to these observations in children, night blindness is not as readily a feature of vitamin A deficiency in adults. In albino rats, Dowling[286] found an initially more marked reduction of the a-wave to threshold stimuli, associated with a decline in rhodopsin concentration of the rod outer segments, eventually followed by a b-wave reduction. Dowling noted a linear relationship between the rhodopsin concentration and the logarithm visual threshold of the ERG. In the early stages of vitamin A deficiency in humans, a reversal of initially reduced ERG amplitudes occurs when parenteral vitamin A is administered. In 1966, Genest[287] reported that both a- and b-waves of the ERG were equally affected by decreased vitamin A blood serum levels in humans. Genest did not note the initial selective a-wave reduction reported by Dowling in his experiments with albino rats. Patients with vitamin A deficiency and night blindness associated with chronic alcoholism and liver cirrhosis do not show the appreciable delays in b-wave implicit times seen in some patients with retinitis pigmentosa.[288] Further, after vitamin A administration, cone function recovers more quickly than rod function in the central retina, compared to the retinal periphery, where rod function recovers more quickly, possibly due to rod–cone rivalry for vitamin A.[288,289]

Retinoids, which are derivatives of vitamin A, are being used for the treatment of dermatologic disorders such as acne and psoriasis. One such derivative, isotretinoin (13-cis-retinoic acid,), was reported by Weleber et al[290] to cause abnormal retinal function, with the complaint of poor night vision and a reduction in ERG amplitudes most apparent under scotopic conditions. A similar reduction, primarily in rod-mediated ERG amplitudes, was reported with the use of fenretinide, a synthetic retinoid.[291]

1-20

OPTIC NERVE AND GANGLION CELL DISEASE

Some authors claim the presence of centrifugal efferent fibers within the optic nerve has an inhibitory effect on the ERG. Thus, apparently as a consequence of the interruption of this inhibitory effect, in some optic nerve diseases the ERG has been reported as supernormal. Other investigators have noted that surgical sectioning of the optic nerve without disturbance of the retinal circulation has no apparent effect on the ERG amplitude.

Since the ganglion cells do not contribute to the flash-elicited full-field ERG response, an essentially normal ERG is obtained in most eyes that are blind as a result of glaucoma. Further, patients with infantile amaurotic familial idiocy (Tay-Sachs disease), a disorder known to involve the retinal ganglion cells, have a normal ERG. Some investigators have noted a lowered critical flicker fusion frequency associated with the retrobulbar neuritis of multiple sclerosis. Further, Fazio et al[292] claim that, if proper gender and age controls are utilized, reduced flash-evoked ERG amplitudes and associated prolonged implicit times can be encountered in eyes with advanced glaucomatous optic nerve damage. Secondary ocular ischemia or vascular pathology, in addition to retrograde degeneration of bipolar and photoreceptor cells (secondary to increased intraocular pressure), was considered a possible explanation.

1-21

OPAQUE LENS OR VITREOUS

Particularly dense lens opacities can reduce the ERG a- and b-wave amplitudes. When present, the amplitude decrease is associated with a comparable increase in implicit time, both relating to the resultant decrease in effective stimulus intensity. It takes a more intense stimulus to reach the same amplitude and implicit time compared to an eye with clear media. Lens opacities do not appear to modify the oscillatory potentials. Patients with mild or possibly even moderate degrees of vitreous hemorrhage generally have normal or only moderately subnormal ERG a- and b-wave amplitudes. As with lens opacities, the implicit times are prolonged because of the apparent decrease in stimulus intensity reaching the photoreceptors. With long-standing vitreous hemorrhages, a marked reduction of ERG amplitudes is always possible because of the ever-present threat of siderotic changes occurring within the retina secondary to blood breakdown products. Fuller et al[293] reported that very high-intensity (bright-flash) elicited ERG recordings could be of value for predicting postvitrectomy improvement in visual function of diabetic patients with vitreous opacities. However, nondetectable preoperative ERG responses have been reported in patients with dense vitreous opacities who showed appreciable improvement in visual function after vitreous surgery.[294,295] Further, results from a bright-flash elicited ERG need to be interpreted with particular caution in patients who have undergone extensive panretinal photocoagulation. Exposure of eyes with clear media to a bright-flash stimulus should be avoided.

1-22

DIABETIC RETINOPATHY

The most frequently reported ERG abnormalities in patients with diabetic retinopathy include a reduction in b-wave amplitude and a reduction or absence of oscillatory potentials. For the most part, reductions in b-wave amplitude are noted in eyes with proliferative retinopathy. These amplitude reductions likely represent a quantitative measure of overall inner retinal ischemia and/or hypoxia. Reduced oscillatory potential amplitudes have been noted at early stages of retinopathy, when ERG a- and b-wave amplitudes are normal,[296] while delayed oscillatory potential implicit times have been reported as an early functional abnormality in eyes with mild or even no retinopathy. Yonemura and Kawasaki[297] have emphasized a selective delay in the implicit time of OPs, in the absence of amplitude reductions, as an early functional change of the retina in diabetic patients. Bresnick and Palta[298] noted a delay in the 30-Hz ERG flicker implicit time, which became progressively more prolonged with increasing severity in retinopathy. These same authors also showed that the summed amplitudes of OPs decrease and implicit times in some of these potentials increase as the severity of diabetic retinopathy increases.[298-300] Bresnick et al[301] reported a 10-fold higher rate of progression to high-risk characteristics of diabetic retinopathy, as defined by the Diabetic Retinopathy Study, in eyes with reduced oscillatory potential amplitudes compared to those eyes with normal amplitudes at study entry.

Later changes in both a- and b-wave amplitudes with more severe cases of diabetic retinopathy are probably related to associated arteriosclerotic changes, vitreous hemorrhages, and frequently to retinal detachments present in the advanced stages of diabetic retinopathy.

Deneault et al[302] used ERG amplitudes and implicit times to document the benefits of multiple-dose daily insulin therapy to conventional once-a-day injection of insulin in streptozotocin-induced diabetic rats.

The effect of panretinal photocoagulation on ERG amplitudes in diabetic patients has been addressed in several publications. Lawwill and O'Connor[303] reported an average 10% reduction in a- and b-wave amplitudes when approximately 20% of the retinal area was photocoagulated. In general, the percentage decrease in ERG amplitude was less than the total area photocoagulated. Wepman et al[304] reported findings on 10 diabetic patients who had argon laser photocoagulation burns to 15–18% of the retina. A mean reduction of 61% in the rod b-wave amplitude, a 35% decrease in the photopic (cone) b-wave response, and a 39% decrease in the combined scotopic rod and cone response to a bright light stimulus were observed. Reduction of the ERG following laser treatment was positively correlated with the total area of retina destroyed, provided that at least 7–10% of the retina had been treated. Patients with larger pretreatment b-wave amplitudes (greater than 300 μV) showed the greatest decrease in ERG amplitude, while patients with smaller pretreatment amplitudes (less than 200 μV) showed the least amplitude change. Therefore, the total area of photocoagulation may be less reliable in consistently predicting the degree of postphotocoagulation reduction in ERG amplitudes than is the size of the ERG amplitude prior to treatment. Those with more reduced pretreatment ERG amplitudes likely have a greater degree of microangiopathy. Photocoagulation of less viable regions of the retina would not be expected to contribute to a reduction in ERG amplitudes, in contrast to destruction of a normal-functioning retina. The disproportionately large percentage reduction in ERG amplitudes compared to the area of photocoagulation observed by Wepman et al[304] contrasts with the findings of Lawwill and O'Connor,[303] but is consistent with the findings of Ogden et al[305] and Frank.[306]

1-23

MISCELLANEOUS CONDITIONS

1-23-1 Retinal Detachment

In general, the amplitudes of both a- and b-waves are related to the degree of retinal detachment. Karpe and Rendahl[307] have reported subnormal ERG values in the normal eye when the contralateral eye has a retinal detachment. In an eye with a detached retina, the ERG deteriorates in relation to both the size of the detached area and the decrease in retinal function. A nondetectable or minimal response implies a total or large detachment, respectively, with a poorly functioning retina. There is, however, some disagreement as to whether these patterns imply a poor prognosis for successful reattachment and ultimate return of visual function. Both Rendahl[308] and Jacobson et al[309] indicated that the type of response is of some prognostic value, while Schmoger[310] and François and de Rouck[311] did not place significant prognostic value on the ERG response.

1-23-2 Thyroid and Other Metabolic Dysfunctions

Several investigators have noted the bioelectrical activity of the retina to be increased in patients with hyperthyroidism and reduced in patients with myxedema. Particularly in thyrotoxic exophthalmos, both a- and b-waves of the ERG were reported as significantly supernormal. Wirth[312] substantiated this finding of supernormal responses and emphasized the generally high correlation with the plasma protein-bound iodine. The ERG is said to be a sensitive test in hyperthyroidism, being supernormal even when the basal metabolic rate and protein-bound iodine may still retain normal values.

Pearlman and Burian[313] noted that the supernormal ERG amplitudes in patients with hyperthyroidism tend to diminish significantly in the course of antithyroid treatment. Conversely, they reported that myxedematous patients with initially subnormal amplitudes showed an increase in ERG response while receiving thyroid treatment. These authors suggested that the ERG may be a reasonably reliable, objective means of estimating the success of medical and surgical therapy in thyroid disease. Wirth[312] stated that the adrenal medulla had no influence on the ERG, noting that patients with a pheochromocytoma, as well as those injected with 1 mg of epinephrine (*Adrenalin*), still had normal b-wave amplitudes. JT Pearlman, MD, and HM Burian, MD (unpublished data), however, recorded a selective b-wave enlargement in patients with an adrenal pheochromocytoma. Patients with Cushing's disease and patients receiving exogenous steroids can manifest a supernormal ERG.[314] Zimmerman et al[315] recorded an increase of approximately 200–300% for both photopic and scotopic a- and b-wave amplitudes in patients receiving 40 mg of prednisone daily for 3 weeks. Negi et al[316] could not demonstrate a similar effect of betamethasone or dexamethasone in their study on albino rabbits. Aldosterone was found by other investigators to increase the ERG b-wave, but decrease the c-wave amplitude in the rabbit.[317] An increase in ERG amplitude in primary aldosteronism was also noted by Wirth and Tota.[314] These investigators also noted that an intravenous injection of aldosterone in rabbits could increase the ERG b-wave amplitude by 45%. A summary of ERG abnormalities in various endocrine and other metabolic disorders is available in a review by Wirth.[318]

1-23-3 Myotonic Dystrophy

Ocular findings in Steinert's myotonic dystrophy include cataract, low intraocular pressure, and, less frequently, macular and/or peripheral pigmentary changes. Burian and Burns[319] obtained ERG data on patients with myotonic dystrophy and noted an average b-wave reduction of 40–45% in these patients, even in those with no or minimal ophthalmoscopically visible changes. A reduction in scotopic b-wave amplitude was also reported by Cavallacci et al,[320] who additionally found a prolonged scotopic b-wave implicit time. Stanescu and Michiels[321] noted a reduction of the photopic as well as scotopic b-wave amplitude in addition to a delayed scotopic b-wave implicit time. The reduction in b-wave amplitude in patients with myotonic dystrophy differs from the normal ERG amplitudes found in patients with myotonia congenita, another autosomal dominant disorder of skeletal muscle.[322]

1-23-4 Albinism

Three general modes of inheritance may be found in albinism: (1) autosomal recessive oculocutaneous, (2) autosomal dominant oculocutaneous, and (3) X-linked recessive or autosomal recessive ocular. The oculocutaneous groups include both tyrosinase-positive and tyrosinase-negative individuals. Krill and Lee[323] found that oculocutaneous and ocular albinos showed an increased ERG response in the dark-adapted state that was greater for the oculocutaneous albinos. Flicker fusion frequencies and light-adapted responses were normal. The responses of female carriers for ocular albinism were not significantly different from those of a control group. The increased scotopic responses tended to become more normal in some older albinotic patients, commensurate with an increase in ocular as well as skin pigmentation. Internally reflected light, because of its lack of absorption by sparse or absent pigment melanin granules, was believed to account for the supernormal scotopic responses. All groups of patients with albinism had normal EOGs. Subsequent studies by Bergsma and Kaiser-Kupfer[324] of a family with autosomal dominant oculocutaneous albinism and by Tomei and Wirth[325] of a group of ocular albinos did not find supernormal ERG amplitudes. Similarly, Wack et al[326] found normal ERG amplitudes in 9 tyrosinase-positive oculocutaneous albinos. Only at high flash luminances, and primarily after dark adaptation, did tyrosinase-negative oculocutaneous albinos show increased ERG amplitudes.

1-23-5 Tay-Sachs Disease

Also initially called amaurotic familial idiocy, this disease was first described clinically by Tay in 1881 and both clinically and pathologically by Sachs in 1887. The disease is a hereditary cerebromacular degeneration associated with lipoidal degeneration of ganglion cells of both the retina and the central nervous system. The disease almost exclusively affects Jewish children of Eastern European origin, with a female predilection of approximately 3 to 1. There is an absence of the enzyme hexosaminidase A in serum, leukocytes, and skin fibroblasts, among other tissues, in patients with this disorder. The characteristic fundus picture is a cherry-red spot within the macula, resulting from the deposition of lipid material in ganglion cells. Optic atrophy is also present. Because cells proximal to the bipolar cell layer are involved, the ERG is typically normal, as noted in Section 1-20.

1-23-6 Congenital Red-Green Color Defects

Congenital red-green color defects can affect either the long-wavelength cones (protan defect) or the middle-wavelength cones (deutan defect). The protan defect can be either partial (protanomaly) or complete (protanopia). A similar classification system exists for the deutan defect, which can be either partial (deuteranomaly) or complete (deuteranopia). The responses to white-light, full-field flash stimuli are normal under both photopic and scotopic levels of adaptation. Of note, the initial portion of the dark-adapted cone-mediated response to orange-red light (x-wave) is generally absent or of minimal amplitude in protanopes or protanomals, but is present in deuteranopes and deuteranomals.

1-23-7 Intraocular Foreign Bodies

The time and rate of ERG changes associated with the retention of an intraocular foreign body depend on the (1) nature of the metal (its alloy content), (2) degree of encapsulation, (3) size of the particle, (4) location, and (5) duration within the eye. Iron and copper affect the ERG rather rapidly. Aluminum only rarely produces a pathologic ERG, and then usually after a relatively long duration. Metallic particles within the lens or anterior segment generally do not cause significant electrophysiologic changes. The inciting particles generally reside within the vitreous, ciliary body, or retina.

The earliest fundus changes noted in patients with retained intraocular iron particles (siderosis bulbi) include a patchy pigmentary degeneration in the retinal periphery with pigment clumping. There are a corresponding depression in peripheral visual fields and elevated thresholds on dark-adaptation testing. The earliest ERG findings in patients with siderosis bulbi from retained iron particles may be a transient supernormal response. With progressive intraocular changes, a negative pattern is eventually followed by a nondetectable response in the most severe cases (Figure 1-39). Caution in initial interpretation is warranted, since a traumatically induced retinal edema from the foreign body may produce a subnormal ERG, with subsequent normalization of the response when the edema has resolved. The ERG changes induced by the foreign body pass through the stages of a transient supernormal response, a negative-positive response, a negative-negative response, and finally nondetectable responses. The ERG changes are reversible in the negative-positive and early negative-negative stages (b-wave amplitude not reduced beyond 50% of the other eye). In the somewhat more advanced negative-negative stage (b-wave amplitude reduced

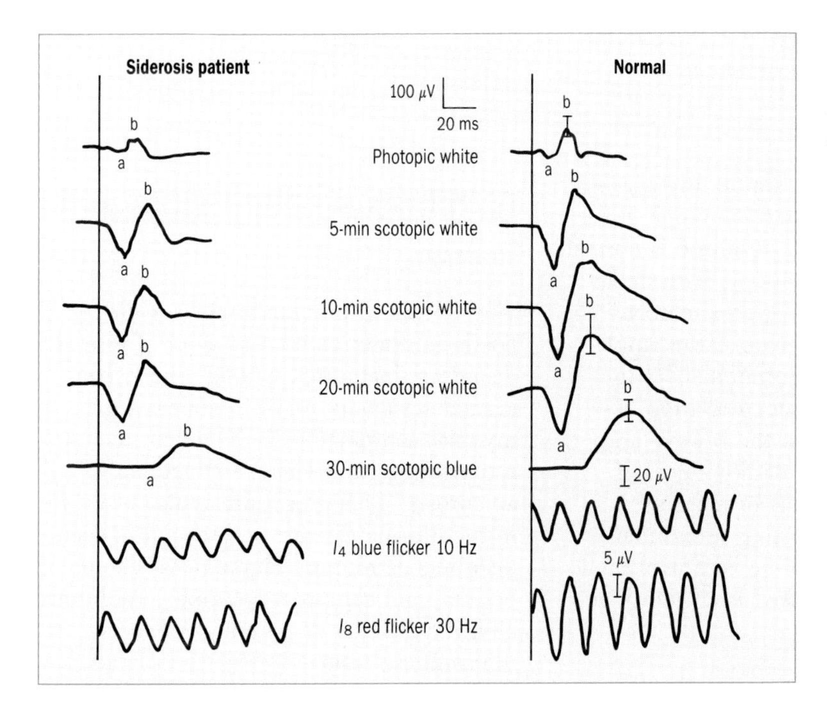

Figure 1-39 *Reduction in ERG amplitude in a patient with siderosis from a retained intraocular foreign body. Reduction in b-wave amplitude is somewhat more apparent than in the a-wave.*

beyond 50–75% of the other eye), it is still possible to stabilize the response or prevent further deterioration by removing the foreign body.[327]

In studies on rabbits, Declercq et al[328] showed that most of the siderotic retinal damage demonstrated by the ERG occurred within the first week. In a subsequent report, Declercq et al[329] demonstrated that foreign body removal after 1 week did not result in normalization of ERG amplitudes in this same rabbit model. Magnetic iron particles have the most toxic effects on the retina. Generally, nonmagnetic iron particles, which are stainless steel, do not affect the ERG as markedly. As noted in Section 1-21, recurrent vitreal hemorrhages from any source may cause retinal damage and corresponding ERG changes similar to those produced by an iron foreign body, likely due to iron liberated from erythrocytes.

1-23-8 Myopia

The ERG is normal in simple myopia, in the absence of degenerative myopic fundus changes. The ERG response is normal in more than 20% of patients with malignant (high) myopia. In approximately 75% of cases of malignant myopia, the amplitude of the b-wave is subnormal.[330] In general, there is a direct correlation between the amplitude of the b-wave and the degree of myopia, with patients of approximately 7 diopters or more of myopia already manifesting subnormal b-wave amplitudes. There is, however, no relationship between the visual acuity and the b-wave amplitude. If the ERG pattern in patients with myopia is either negative or markedly reduced in amplitude, the myopia may be associated with an inherited retinal disorder, ie, congenital stationary night blindness or retinitis pigmentosa.

1-23-9 Taurine Deficiency

Taurine is an amino acid abundantly concentrated in the outer segments of rods and cones. Kittens fed a casein diet develop a selective decrease in plasma and retinal taurine levels, resulting in a retinal photoreceptor cell degeneration and associated reduction of ERG amplitudes.[331,332]

In children receiving long-term parenteral nutrition, reductions in plasma taurine levels can be accompanied by generally mild and reversible ERG abnormalities, including primarily prolonged implicit times and small amplitude reductions.[333,334]

1-24

SUMMARY

The ERG can provide important diagnostic information on a number of retinal disorders, including, among others, congenital stationary night blindness with myopia, achromatopsia, X-linked juvenile retinoschisis, and Leber's congenital amaurosis. This procedure is also of value in monitoring disease progression in such disorders as retinitis pigmentosa, choroideremia, and cone-rod dystrophy. Additionally, the ERG is of potential value in determining retinal toxicity associated with the use of such drugs as thioridazine or in vitamin A deficiency as well as for ascertaining possible retinal toxicity from a retained intraocular foreign body. For the ERG to provide reliable information, it is vital for each investigator to comprehensively establish a normative database and to maintain both stimulus and background light sources appropriately calibrated. This statement also applies to foveal cone ERG and pattern ERG measurements, which provide potentially useful information about foveal function and more proximal retinal cell function respectively.

REFERENCES

1. Dewar J: The physiologic action of light. *Nature* 1877;**15**:433–435.
2. Riggs LA: Continuous and reproducible records of the electrical activity of the human retina. *Proc Soc Exp Biol Med* 1941;**48**:204–207.
3. Karpe G: The basis of clinical electroretinography. *Acta Ophthalmol* 1945(suppl 24): 1–118.
4. Einthoven W, Jolly W: The form and magnitude of the electrical response of the eye to stimulation by light at various intensities. *Quart J Exp Physiol* 1908;**1**:373–416.
5. Granit R: The components of the retinal action potential and their relation to the discharge in the optic nerve. *J Physiol* 1933;**77**: 207–240.
6. Noell WK: The effect of iodoacetate on the vertebrate retina. *J Cell Physiol* 1951;**37**: 283–307.
7. Dowling JE: Organization of vertebrate retinas. *Invest Ophthalmol* 1970;**9**:655–680.
8. Kline RP, Ripps H, Dowling JE: Generation of b-wave currents in the skate retina. *Proc Natl Acad Sci USA* 1978;**75**:5727–5731.
9. Newman EA, Odette LL: Model of electroretinogram b-wave generation: a test of the K^+ hypothesis. *J Neurophysiol* 1984;**51**:164–182.
10. Ripps H, Witkovsky P: Neuron-glia interaction in the brain and retina. In: Osborne NN, Chader GJ, eds. *Progress in Retinal Research.* Elmsford, NY: Pergamon Press; 1985;**4**:181–219.
11. Stockton RA, Slaughter MM: B-wave of the electroretinogram: a reflection of bipolar cell activity. *J Gen Physiol* 1989;**93**:101–122.
12. Karwoski CJ, Proenza LM: Relationship between Müller cell responses: a local transretinal potential and potassium flux. *J Neurophysiol* 1977;**40**:244–259.
13. Marmor MF, Lurie M: Light-induced electrical responses of the retinal pigment epithelium. In: Zinn KM, Marmor MF, eds. *The Retinal Pigment Epithelium.* Cambridge, Mass: Harvard University Press; 1979:226–244.
14. Arden GB, Tansley K: The spectral sensitivity of the pure-cone retina of the grey squirrel. *J Physiol (Lond)* 1955;**127**:592–602.
15. Tomita T, Matsuura T, Fujimoto M, et al: The electroretinographic c- and e-waves with special reference to the receptor potential. Sixteenth ISCEV Symposium. *Jpn J Ophthalmol Proc Ser* 1979:15–25.
16. Marmor MF, Hock PA: A practical method for c-wave recording in man. *Doc Ophthalmol Proc Ser* 1982;**31**:67–72.
17. Kawasaki K, Yonemura D, Nakazato H, et al: Analysis of rapid off-response in electroretinogram major participation of receptor potential. *Acta Soc Ophthalmol Jpn* 1980;**84**: 1574–1580.
18. Nagata M: Studies on the photopic ERG of the human retina. *Jpn J Ophthalmol* 1963;**7**: 96–124.

19. Jacobson JH, Gestring GF: Centrifugal influence on the electoretinogram. *Arch Ophthalmol* 1958;60:295–302.

20. Gills JP: The electroretinogram after section of the optic nerve in man. *Am J Ophthalmol* 1966;**62**:287–291.

21. Horsten GPM, Winkelman JE: Effect of section of the optic nerve on histologic differentiation and electrical activity of the retina in dogs. *Ophthalmologica* 1969;**157**:293–300.

22. Armington JC, Tepas DI, Kropfl WJ, et al: Summation of retinal potentials. *J Opt Soc Am* 1980;**51**:877–886.

23. François J, de Rouck A: Behavior of ERG and EOG in localized retinal destruction by photocoagulation. In: Burian HM, Jacobson H, eds. *Clinical Electroretinography*. Proceedings of the Third International Symposium, 1964. Elmsford, NY: Pergamon Press; 1966:191–202.

24. Schuurmans RP, van Lith GHM, Oosterhuis JA: Photocoagulation and the electroretinogram. *Doc Ophthalmol Proc Ser* 1978;**15**: 297–301.

25. Brown KT, Murakami M: A new receptor potential of the monkey retina with no detectable latency. *Nature* 1964;**201**:626–628.

26. Pak WL, Cone RA: Isolation and identification of the initial peak of the early receptor potential. *Nature* 1965;**204**:836–838.

27. Goldstein EB, Berson EL: Rod and cone contributions to the human early receptor potential. *Vision Res* 1970;**10**:207–218.

28. Sieving PA, Fishman GA: Rod contribution to the human early receptor potential (ERP) estimated from monochromats' data. *Doc Ophthalmol Proc Ser* 1982;**31**:95–102.

29. Cobb WA, Morton HB: A new component of the human electroretinogram. *J Physiol* 1954;**123**:36P–37P.

30. Yonemura D: The oscillatory potential of the electroretinogram. *Acta Soc Ophthalmol Jpn* 1962;**66**:1566–1584.

31. Speros P, Price J: Oscillatory potentials: history, techniques and potential use in the evaluation of disturbances of the retinal circulation. *Surv Ophthalmol* 1981;**25**:237–252.

32. Peachey NS, Alexander KR, Fishman GA: Rod and cone system contributions to oscillatory potentials: an explanation for the conditioning flash effect. *Vision Res* 1987;**27**: 859–866.

33. Korol S, Leuenberger PM, Englert U, et al: In vivo effects of glycine on retinal ultrastructure and averaged electroretinograms. *Brain Res* 1975;**97**:235–251.

34. Wachtmeister L: Further studies of the chemical sensitivity of the oscillatory potentials of the electroretinogram (ERG), I: GABA and glycine antagonists. *Acta Ophthalmol* 1980;**58**: 712–725.

35. Wachtmeister L: Further studies of the chemical sensitivity of the oscillatory potentials of the electroretinogram (ERG), II: glutamate, aspartate- and dopamine antagonists. *Acta Ophthalmol* 1981;**59**:247–258.

36. Hirose T, Wolf E, Hara A: Electrophysiological and psychophysical studies in congenital retinoschisis of X-linked recessive inheritance. *Doc Ophthalmol Proc Ser* 1977;**13**:173–184.

37. Miyake Y, Yagasaki K, Horiguchi M, et al: Congenital stationary night blindness with negative electroretinogram: a new classification. *Arch Ophthalmol* 1986;**104**:1013–1020.

38. Kubota Y, Kubota S: ERG of Behcet's disease and its diagnostic significance. *Doc Ophthalmol Proc Ser* 1980;**23**:91–93.

39. Berson EL, Sandberg MA, Rosner B, et al: Natural cause of retinitis pigmentosa over a three-year interval. *Am J Ophthalmol* 1985;**99**: 240–251.

40. Burian HM, Allen L: A speculum contact lens electrode for electroretinography. *EEG Clin Neurophysiol* 1954;6:509–511.

41. Lawwill T, Burian HM: A modification of the Burian-Allen contact-lens electrode for human electroretinography. *Am J Ophthalmol* 1966;6:1506–1509.

42. Dawson WW, Trick GL, Litzkow CA: Improved electrode for electroretinography. *Invest Ophthalmol Vis Sci* 1979;**18**:988–991.

43. Dawson WW, Trick GL, Maida TM: Evaluation of the DTL corneal electrode. *Doc Ophthalmol Proc Ser* 1982;**31**:81–88.

44. Grounauer PA: The new single use ERG corneal contact lens electrode and its clinical application. *Doc Ophthalmol Proc Ser* 1982;**31**: 89–93.

45. Chase WW, Fradkin NE, Tsuda S: A new electrode for electroretinography. *Am J Optom Physiol Optics* 1976;**53**:668–671.

46. Arden GB, Carter RM, Hogg C, et al: A gold foil electrode extending the horizons for clinical electroretinography. *Invest Ophthalmol Vis Sci* 1979;**18**:421–426.

47. Borda BP, Gilliam RM, Coats AC: Gold-coated Mylar (GCM) electrode for electroretinography. *Doc Ophthalmol Proc Ser* 1978;**15**: 339–343.

48. Sierpinski-Bart J, Moran A, Hocherman S, et al: Non-corneal electroretinography for pediatric ophthalmology. *Metab Ophthalmol* 1978;**2**:387–388.

49. van Norren D: The technical limitations of clinical electroretinography. *Doc Ophthalmol Proc Ser* 1982;**31**:3–12.

50. Sieving PA, Fishman GA, Maggiano J: Corneal wick electrode for recording bright flash electroretinograms and early receptor potential. *Arch Ophthalmol* 1978;96:899–900.

51. Weleber RG, Eisner A: Retinal function and physiological studies. In: Newsome DA, ed. *Retinal Dystrophies and Degenerations*. New York: Raven Press; 1988:21–69.

52. Burian HM: Electric responses of the human visual system. *Arch Ophthalmol* 1954;**51**:509–524.

53. Armington JC, Biersdorf WR: Long-term light adaptation of the human electroretinogram. *J Comp Physiol Psychol* 1958;**51**:1–5.

54. Gouras P, MacKay CJ: Growth amplitude of the human cone electroretinogram with light adaptation. *Invest Ophthalmol Vis Sci* 1989;**30**: 625–630.

55. Lachapelle P: Analysis of the photopic electroretinogram recorded before and after dark adaptation. *Can J Ophthalmol* 1987;**22**: 354–361.

56. Peachey NS, Alexander KR, Fishman GA, et al: Properties of the human cone system electroretinogram during light adaptation. *Appl Optics* 1989;**28**:1145–1150.

57. Miyake Y, Horiguchi M, Ota I, et al: Adaptational change in cone-mediated electroretinogram in human and carp. *Neurosci Res* 1988;8(suppl):S1–S13.

58. Sandberg MA, Berson EL, Effron MH: Rod-cone interaction in the distal human retina. *Science* 1981;**212**:829–831.

59. Brunette JR: The human electroretinogram during dark adaptation. *Arch Ophthalmol* 1969;**82**:491–498.

60. Johnson MA, Massof RW: The photomyoclonic reflex: an artifact in the clinical electroretinogram. *Br J Ophthalmol* 1982;66:368–378.

61. Massof RW, Wu L, Finkelstein D, et al: Properties of electroretinographic intensity-response functions in retinitis pigmentosa. *Doc Ophthalmol* 1981;**57**:279–296.

62. Motokawa K, Mita T: Uber eine einfachere Untersuchungsmethode und Eigenschaften der Aktionsstrome der netzhaut des Menschen. *Tohoku J Exp Med* 1942;**42**:114–133.

63. François J, Verriest G, de Rouck A: Pathology of the x-wave of the human electroretinogram. *Br J Ophthalmol* 1956;**40**:439–443.

64. Müller-Limmroth W: Der Einfluss der Dauer des Reizlichtes auf der Electroretinogramm. *Arch Gesamte Physiol* 1953;**257**:35–47.

65. Johnson E, Bartlett N: Effect of stimulus duration on electrical responses of the human electroretinogram. *Br J Ophthalmol* 1956;**40**: 439–443.

66. Gouras P: Electroretinography: some basic principles. *Invest Ophthalmol* 1970;**9**:557–569.

67. Rabin AR, Berson EL: A full-field system for clinical electroretinography. *Arch Ophthalmol* 1974;**92**:59–63.

68. Kooijman AC: Light distribution on the retina of a wide-angle theoretical eye. *J Opt Soc Am* 1983;**73**:1544–1550.

69. Mahneke A: Electroretinography with double flashes. *Acta Ophthalmol* 1957;**35**:131–141.

70. Henkes HE: The use of electroretinography in measuring the effects of vasodilation. *Angiology* 1951;**2**:125–131.

71. Horsten GP, Winkelman JE: The electric activity of the eye in the first days of life. *Acta Physiol Pharmacol Neerl* 1965;**13**:1.

72. Zetterstrom B: The electroretinogram of the newborn infant. In: Wirth A, ed. *Proceedings of the Eighth ISCERG Symposium*. Pisa: Pacini Mariotti; 1972:1–9.

73. Fulton AB, Hansen RM: Background adaptation in human infants: analysis of b-wave responses. *Doc Ophthalmol Proc Ser* 1982;**31**:191–197.

74. Fulton AB: The development of scotopic retinal function in human infants. *Doc Ophthalmol* 1988;**69**:101–109.

75. Fulton AB, Hansen RM: Electroretinography: application to clinical studies of infants. *J Pediatr Ophthalmol Strab* 1985;**6**:251–255.

76. Peterson H: The normal b-potential in the single-flash clinical electroretinogram. *Acta Ophthalmol* 1968(suppl 99):1–77.

77. Vainio-Mattila B: The clinical electroretinogram, II: the difference between the electroretinogram in men and in women. *Acta Ophthalmol* 1951;**29**:25–32.

78. Zeidler I: The clinical electroretinogram, IX: the normal electroretinogram—value of the b-potential in different age groups and its differences in men and women. *Acta Ophthalmol* 1959;**37**:294–301.

79. Karpe G, Rickenbach K, Thomasson S: The clinical electroretinogram, I: the normal electroretinogram above fifty years of age. *Acta Ophthalmol* 1950;**28**:301–305.

80. Martin DA, Heckenlively JR: The normal electroretinogram. *Doc Ophthalmol Proc Ser* 1982;**31**:135–144.

81. Weleber RG: The effect of age on human cone and rod Ganzfeld electroretinograms. *Invest Ophthalmol Vis Sci* 1981;**20**:392–399.

82. Lehnert W, Wunsche H: Das Elektroretinogramm in verschiedenen Lebensaltern. *Graefes Arch Klin Exp Ophthalmol* 1966;**170**:147–155.

83. Pallin O: The influence of the axial length of the eye on the size of the recorded b-potential in the clinical single-flash electroretinogram. *Acta Ophthalmol* 1969(suppl 101):1–57.

84. Padmos P, van Norren D: Cone pigment regeneration: the influence of halothane anesthesia. *Doc Ophthalmol Proc Ser* 1975;**11**: 145–148.

85. van Norren D, Padmos P: Cone dark adaptation: the influence of halothane anesthesia. *Invest Ophthalmol* 1975;**14**:212–227.

86. van Norren D, Padmos P: The influence of various inhalation anesthetics on dark adaptation. *Doc Ophthalmol Proc Ser* 1975;**11**:149–151.

87. van Norren D, Padmos P: Halothane retards dark adaptation. *Doc Ophthalmol Proc Ser* 1974;**4**:155–159.

88. Birch DG, Berson EL, Sandberg MA: Diurnal rhythm in the human rod ERG. *Invest Ophthalmol Vis Sci* 1984;**25**:236–238.

89. Riggs LA, Johnson EP, Shick AML: Electrical responses of the human eye to moving stimulus patterns. *Science* 1964;**144**:567.

90. Sandberg MA, Ariel M: A hand-held, two-channel stimulator-ophthalmoscope. *Arch Ophthalmol* 1977;**95**:1881–1882.

91. Birch DG, Fish GE: Focal cone electroretinograms: aging and macular disease. *Doc Ophthalmol* 1988;**69**:211–220.

92. Sandberg MA, Jacobson SG, Berson EL: Foveal cone electroretinograms in retinitis pigmentosa and juvenile macular degeneration. *Am J Ophthalmol* 1979;**88**:702–707.

93. Jacobson SG, Sandberg MA, Effron MH, et al: Foveal cone electroretinograms in strabismic amblyopia: comparison with juvenile macular degeneration, macular scars and optic atrophy. *Trans Ophthalmol Soc UK* 1980;**99**: 353–356.

94. Birch DG, Jost BF, Fish GE: The focal electroretinogram in fellow eyes of patients with idiopathic macular holes. *Arch Ophthalmol* 1988;**106**:1558–1563.

95. Fish GE, Birch DG: The focal electroretinogram in the clinical assessment of macular disease. *Ophthalmology* 1989;**96**:109–114.

96. Salzman J, Seiple W, Carr R, et al: Electrophysiologic assessment of aphakic cystoid macular oedema. *Br J Ophthalmol* 1986;**70**: 819–824.

97. Miyake Y, Shiroyama N, Ota I, et al: Local macular electroretinographic responses in idiopathic central serous chorioretinopathy. *Am J Ophthalmol* 1988;**106**:546–550.

98. Sokol S, Nadler D: Simultaneous electroretinograms and visual evoked potentials from adult amblyopes in response to a pattern stimulus. *Invest Ophthalmol Vis Sci* 1979;**18**: 848–855.

99. Kaufman D, Celesia GG: Simultaneous recording of pattern electroretinogram and visual evoked response in neuro-ophthalmologic disorders. *Neurology* 1985;**35**:644–651.

100. Rimmer S, Katz B: The pattern electroretinogram: technical aspects and clinical significance. *J Clin Neurophysiol* 1989;6:85–99.

101. Odom J, Dawson W, Romano P, et al: Human pattern evoked retinal response (PERR): spatial tuning and development. *Doc Ophthalmol Proc Ser* 1983;**37**:265–271.

102. Berson EL: Electrical phenomena in the retina. In: Moses RA, Hart WM Jr, eds. *Adler's Physiology of the Eye: Clinical Application*. 8th ed. St Louis, CV Mosby Co; 1987:506–567.

103. Sokol S, Jones K, Nadler D: Comparison of the spatial response properties of the human retina and cortex as measured by simultaneously recorded pattern ERGs and VEPs. *Vision Res* 1983;**23**:723–727.

104. Hess RF, Baker CL Jr: Human pattern-evoked electroretinogram. *J Neurophysiol* 1984; **51**:939–951.

105. Marx M, Ghilardi M, Bodis-Wollner I: Refractive error and the pattern ERG and VEP in alert normal and aphakic monkeys. *Invest Ophthalmol Vis Sci* 1986;**26**(suppl):281.

106. Millodot M, Riggs LA: Refraction determined electrophysiologically. *Arch Ophthalmol* 1970;**84**:272–278.

107. Siegel MJ, Marx MJ, Bodis-Wollner I, et al: The effect of refractive error on pattern-electroretinogram in primates. *Curr Eye Res* 1986;**5**:183–187.

108. Maffei L, Fiorentini A: Electroretinographic responses to alternating gratings before and after transection of the optic nerve. *Science* 1981;**211**:953–955.

109. Maffei L, Fiorentini A, Bisti S, et al: Pattern ERG in the monkey after section of the optic nerve. *Exp Brain Res* 1985;**59**:423–425.

110. Bobak P, Bodis-Wollner I, Harnois C, et al: Pattern electroretinograms and visual evoked potentials in glaucoma and multiple sclerosis. *Am J Ophthalmol* 1983;**96**:72–83.

111. van den Berg T, Riemslag F, De Vos G, et al: Pattern ERG and glaucomatous visual field defects. *Doc Ophthalmol* 1986;**61**:335–341.

112. Ringens PJ, Vijfvinkel-Bruinenga S, van Lith GHM: The pattern-elicited electroretinogram, I: a tool in the early detection of glaucoma? *Ophthalmologica* 1986;**192**:171–175.

113. Weinstein GW, Arden GB, Hitchings RA, et al: The pattern electroretinogram (PERG) in ocular hypertension and glaucoma. *Arch Ophthalmol* 1988;**106**:923–928.

114. Trick G: Pattern reversal retinal potentials in ocular hypertensives at high and low risk of developing glaucoma. *Doc Ophthalmol* 1987; **65**:79–85.

115. Bach M, Hiss P, Rover J: Check-size specific changes of pattern electroretinogram in patients with early open-angle glaucoma. *Doc Ophthalmol* 1988;**69**:315–322.

116. Persson H, Wanger P: Pattern-reversal electroretinograms and visual evoked potentials in multiple sclerosis. *Br J Ophthalmol* 1984;**68**: 760–764.

117. Plant G, Hess R, Thomas S: The pattern evoked electroretinogram in optic neuritis: a combined psychophysical and electrophysiological study. *Brain* 1986;**109**:469–490.

118. Kirkham T, Coupland S: The pattern electroretinogram in optic nerve demyelination. *Can J Neurol Sci* 1983;**10**:256–260.

119. Arden GB, Vaegan, Hogg CR: Clinical and experimental evidence that the pattern electroretinogram (PERG) is generated in more proximal retinal layers than the focal electroretinogram (FERG). *Ann NY Acad Sci* 1982; **388**:580–601.

120. Dawson W, Maida T, Rubin M: Human pattern-evoked retinal responses are altered by optic atrophy. *Invest Ophthalmol Vis Sci* 1982;**22**: 796–803.

121. Holder GE: Significance of abnormal pattern electroretinography in anterior visual pathway dysfunction. *Br J Ophthalmol* 1987;**71**: 166–171.

122. Ryan S, Arden GB: Electrophysiological discrimination between retinal and optic nerve disorders. *Doc Ophthalmol* 1988;**68**:247–255.

123. Arden GB, Vaegan, Hogg CR, et al: Pattern ERGs are abnormal in many amblyopes. *Trans Ophthalmol Soc UK* 1980;**100**:453–460.

124. Sokol S: Pattern elicited ERGs and VECPs in amblyopia and infant vision. In: Armington J, Krauskopf J, Wooten B, eds. *Visual Psychophysics and Electrophysiology*. New York: Academic Press; 1978:453–462.

125. Berninger T, Schuurmans RP: Spatial tuning of the pattern ERG across temporal frequency. *Doc Ophthalmol* 1985;**61**:17–25.

126. Schuurmans RP, Berninger T: Luminance and contrast responses in man and cat. *Doc Ophthalmol* 1985;**59**:187–197.

127. Wanger P, Persson HE: Pattern-reversal electroretinograms in ocular hypertension. *Doc Ophthalmol* 1985;**61**:27–31.

128. Arden GB, Hamilton AMP, Wilson-Holt J, et al: Pattern electroretinograms become abnormal when background diabetic retinopathy deteriorates to a preproliferative stage: possible use as a screening test. *Br J Ophthalmol* 1986;**70**:330–335.

129. Coupland SG: A comparison of oscillatory potential and pattern electroretinogram measures in diabetic retinopathy. *Doc Ophthalmol* 1987;**66**:207–218.

130. Henkes HE: Electroretinography in circulatory disturbances of the retina; I: electroretinogram in cases of occlusion of central retinal vein or one of its branches. *Arch Ophthalmol* 1953;**49**:190–201.

131. van Lith G: Subnormal and absent ERGs: what do we mean by these terms? *Doc Ophthalmol Proc Ser* 1982;**31**:13–17.

132. Fishman GA, Farber MD, Derlacki DJ: X-linked retinitis pigmentosa: profile of clinical findings. *Arch Ophthalmol* 1988;**106**:369–375.

133. Fishman GA, Alexander KR, Anderson RJ: Autosomal dominant retinitis pigmentosa: a method of classification. *Arch Ophthalmol* 1985; **103**:366–374.

134. Massof RW, Finkelstein D: Two forms of autosomal dominant primary retinitis pigmentosa. *Doc Ophthalmol* 1981;**51**:289–346.

135. Massof RW, Finkelstein D: Subclassification of retinitis pigmentosa from two-color scotopic static perimetry. *Doc Ophthalmol Proc Ser* 1981;**26**:219–225.

136. Arden GB, Carter RM, Hogg CR, et al: Rod and cone activity in patients with dominantly inherited retinitis pigmentosa: comparisons between psychophysical and electroretinographic measurements. *Br J Ophthalmol* 1983;**67**:405–418.

137. Massof RW, Finkelstein D: Rod sensitivity relative to cone sensitivity in retinitis pigmentosa. *Invest Ophthalmol Vis Sci* 1979;**18**: 263–272.

138. Berson EL, Gouras P, Gunkel RD, et al: Rod and cone responses in sex-linked retinitis pigmentosa. *Arch Ophthalmol* 1969;**81**:215–225.

139. Berson EL, Gouras P, Gunkel RD, et al: Dominant retinitis pigmentosa and reduced penetrance. *Arch Ophthalmol* 1969;**81**:226–234.

140. Marmor MF: The electroretinogram in retinitis pigmentosa. *Arch Ophthalmol* 1979;**97**: 1300–1304.

141. Rothberg D, Weinstein G, Hobson R, et al: Electroretinography and retinitis pigmentosa. *Arch Ophthalmol* 1982;**100**:1422–1426.

142. Sandberg MA, Sullivan PL, Berson EL: Temporal aspects of the dark-adapted cone a-wave in retinitis pigmentosa. *Invest Ophthalmol Vis Sci* 1981;**21**:765–769.

143. Berson EL, Gouras P, Gunkel RD: Rod responses in retinitis pigmentosa dominantly inherited. *Arch Ophthalmol* 1968;**80**:58–67.

144. Berson EL, Kanters L: Cone and rod responses in a family with recessively inherited retinitis pigmentosa. *Arch Ophthalmol* 1970;**84**: 288–297.

145. Berson EL, Goldstein BE: Early receptor potential in dominantly inherited retinitis pigmentosa. *Arch Ophthalmol* 1970;**83**:412–420.

146. Fishman GA, Weinberg AB, McMahon TT: X-linked recessive retinitis pigmentosa: clinical characteristics of carriers. *Arch Ophthalmol* 1986;**104**:1329–1335.

147. Peachey NS, Fishman GA, Derlacki DJ, et al: Rod and cone dysfunction in carriers of X-linked retinitis pigmentosa. *Ophthalmology* 1988;**95**:677–685.

148. Fishman GA, Kumar A, Joseph MD, et al: Usher's syndrome: ophthalmic and neuro-otologic findings suggesting genetic heterogeneity. *Arch Ophthalmol* 1983;**101**:1367–1374.

149. Fishman GA: Hereditary retinal and choroidal diseases: electroretinogram and electro-oculogram findings. In: Peyman GA, Sanders DR, Goldberg MF, eds. *Principles and Practice of Ophthalmology*. Philadelphia: WB Saunders Co; 1980;**2**:857–904.

150. Berson EL: Nutritional and retinal degenerations: vitamin A, taurine, ornithine, and phytanic acid. *Retina* 1982;**2**:236–255.

151. Carr RE, Siegel IM: *Visual Electrodiagnostic Testing: A Practical Guide for the Clinician*. Baltimore, Md: Williams & Wilkins; 1982: 81–84.

152. Pagon RA: Retinitis pigmentosa. *Surv Ophthalmol* 1988;**33**:137–177.

153. Lamy M, Frezal J, Polonovski J, et al: Congenital absence of beta-lipoproteins. *Pediatrics* 1963;**31**:277–289.

154. Berson EL, Howard J: Temporal aspects of the electroretinogram in sector retinitis pigmentosa. *Arch Ophthalmol* 1971;**86**:653–665.

155. Berson EL: Retinitis pigmentosa and allied diseases: applications of electroretinographic testing. *Int Ophthalmol* 1981;**4**:7–22.

156. Carr RE: Vitamin A therapy may reverse degenerative retinal syndrome. *Clin Trends* 1970;**8**:8.

157. Gouras P, Carr RE, Gunkel RD: Retinitis pigmentosa in abetalipoproteinemia: effects of vitamin A. *Invest Ophthalmol* 1971;**10**:784–793.

158. Sperling MA, Hiles DA, Kennerdell JS: Electroretinographic responses following vitamin A therapy in A-beta-lipoproteinemia. *Am J Ophthalmol* 1972;**73**:342–351.

159. Muller DPR, Lloyd JK, Bird AC: Long term management of abetalipoproteinaemia: possible role of vitamin E. *Arch Dis Child* 1977; **52**:209.

160. Azizi E, Zaidman JL, Eshchar J, et al: Case report: abetalipoproteinemia treated with parenteral and oral vitamins A and E and with medium chain triglycerides. *Acta Paediatr Scand* 1978;**67**:797–801.

161. Gills JP, Hobson R, Hanley B, et al: Electroretinography and fundus oculi findings in Hurler's disease and allied mucopolysaccharidoses. *Arch Ophthalmol* 1965;**74**:596–603.

162. Leung LSE, Weinstein GW, Hobson RR: Further electroretinographic studies of patients with mucopolysaccharidoses. *Birth Defects* 1971; **7**:32–40.

163. Bateman JB, Philippart M: Ocular features of the Hagberg-Santavuori syndrome. *Am J Ophthalmol* 1986;**102**:262–271.

164. Jaben S, Flynn JT: Neuronal ceroid lipofuscinosis (Batten-Vogt's disease). In: Smith JL, ed. *Neuro-ophthalmology Update*. New York: Masson Publishers; 1977:299–317.

165. Pinckers A, Bolmers D: Neuronal ceroid lipofuscinosis (E.R.G. et E.O.G.). *Ann Ocul (Paris)* 1974;**207**:523–529.

166. Gottlob I, Leipert KP, Kohlschutter A, et al: Electrophysiologic findings of neuronal ceroid lipofuscinosis in heterozygotes. *Graefes Arch Klin Exp Ophthalmol* 1988;**226**:516–521.

167. Krill AE: *Hereditary Retinal and Choroidal Diseases*. New York: Harper & Row; 1977;**2**: 739–824.

168. Henkes HE: Does unilateral retinitis pigmentosa really exist? An ERG and EOG study of the fellow eye. In: Burian HM, Jacobson JH, eds. *Clinical Electroretinography*. Proceedings of the Third International Symposium, 1964. Elmsford, NY: Pergamon Press; 1966:327–350.

169. Carr RE, Siegel IM: Unilateral retinitis pigmentosa. *Arch Ophthamol* 1973;**90**:21–26.

170. Kandori F, Tamai A, Watanabe T, et al: Unilateral pigmentary degeneration of the retina. *Am J Ophthalmol* 1968;**66**:1091–1101.

171. Kolb H, Galloway NR: Three cases of unilateral pigmentary degeneration. *Br J Ophthalmol* 1964;**48**:471–479.

172. Massof RW, Finkelstein D: Vision threshold profiles in sector retinitis pigmentosa. *Arch Ophthalmol* 1979;**97**:1899–1904.

173. Margolis S, Scher BM, Carr RE: Macular colobomas in Leber's congenital amaurosis. *Am J Ophthalmol* 1977;**83**:27–42.

174. Ripps H, Noble KG, Greenstein VC, et al: Progressive cone dystrophy. *Ophthalmology* 1987;**94**:1401–1409.

175. Krauss HR, Heckenlively JR: Visual field changes in cone-rod degenerations. *Arch Ophthalmol* 1982;**100**:1784–1790.

176. Heckenlively JR, Martin DA, Rosales TO: Telangiectasia and optic atrophy in

cone-rod degenerations. *Arch Ophthalmol* 1981;**99**:1983–1991.

177. Noble KG, Siegel IM, Carr RE: Progressive peripheral cone dysfunction. *Am J Ophthalmol* 1988;**106**:557–560.

178. Krill AE, Deutman AF, Fishman M: The cone degenerations. *Doc Ophthalmol* 1973;**35**: 1–80.

179. Ikeda H, Ripps H: The electroretinogram of the cone-monochromat. *Arch Ophthalmol* 1966;**75**:513–517.

180. Price MJ, Judisch GF, Thompson HS: X-linked congenital stationary night blindness with myopia and nystagmus without clinical complaints of nyctalopia. *J Pediatr Ophthalmol Strab* 1988;**25**:33–36.

181. Heckenlively JR, Martin DA, Rosenbaum AL: Loss of electroretinographic oscillatory potentials, optic atrophy and dysplasia in congenital stationary night blindness. *Am J Ophthalmol* 1983;**96**:526–534.

182. Krill AE, Martin D: Photopic abnormalities in congenital stationary night blindness. *Invest Ophthalmol* 1971;**10**:625–636.

183. Hill DA, Arbel KF, Berson EL: Cone electroretinograms in congenital nyctalopia with myopia. *Am J Ophthalmol* 1974;**78**: 127–136.

184. Siegel IM, Greenstein VC, Seiple WH, et al: Cone function in congenital nyctalopia. *Doc Ophthalmol* 1987;**63**:307–318.

185. Carr RE, Ripps H, Siegel IM, et al: Rhodopsin and the electrical activity of the retina in congenital night blindness. *Invest Ophthalmol* 1966;**5**:497–507.

186. Miyake Y, Horiguchi M, Ota I, et al: Characteristic ERG flicker anomaly in incomplete congenital stationary night blindness. *Invest Ophthalmol Vis Sci* 1987;**11**:1816–1823.

187. Miyake Y, Kawase Y: Reduced amplitude of oscillatory potentials in female carriers of X-linked recessive congenital stationary night blindness. *Am J Ophthalmol* 1984;**98**:208–215.

188. Ripps H, Carr RE, Siegel IM, et al: Functional abnormalities in vincristine-induced night blindness. *Invest Ophthalmol Vis Sci* 1984;**25**:787–794.

189. Berson EL, Lessell S: Paraneoplastic night blindness with malignant melanoma. *Am J Ophthalmol* 1988;**106**:307–311.

190. Ripps H, Siegel IM, Mehaffey L III: The cellular basis of visual dysfunction in hereditary retinal disorders. In: Scheffield JB, Hilfer SR, eds. *Cell and Developmental Biology of the Eye: Heredity and Visual Development.* New York: Springer-Verlag; 1984:171–204.

191. Franceschetti A, François J, Babel J: *Les Hérédo-Dégénérescences Chorio-Rétiniennes.* Paris: Masson et Cie; 1963;**2**:1205–1250.

192. Kubota Y: The oscillatory potentials of the ERG in Oguchi's disease. *Doc Ophthalmol Proc Ser* 1974;**10**:317–324.

193. Carr RE, Ripps H: Rhodopsin kinetics and rod adaptation in Oguchi's disease. *Invest Ophthalmol* 1967;**6**:426–436.

194. Carr RE, Ripps H, Siegel IM: Visual pigment kinetics and adaptation in fundus albipunctatus. *Doc Ophthalmol Proc Ser* 1974;**9**: 193–199.

195. Klien BA, Krill AE: Fundus flavimaculatus: clinical, functional and histopathologic observations. *Am J Ophthalmol* 1967;**64**:3–23.

196. Fishman GA: Hereditary progressive macular dystrophies. In: Smith JL, ed. *Neuro-ophthalmology Update.* New York: Masson Publishers; 1977:73–89.

197. Franceschetti A: Ueber tapeto-retinale Degenerationen im Kindesalter. In Sautter H, ed. *Entwicklung und Fortschritt in der Augenheilkunde.* Stuttgart: Enke Verlag; 1963:107–120.

198. Fishman GA: Fundus flavimaculatus: a clinical classification. *Arch Ophthalmol* 1976; **94**:2061–2067.

199. Fishman GA: Electrophysiology and inherited retinal disorders. *Doc Ophthalmol* 1985;**60**:107–119.

200. Fishman GA: Electroretinography and inherited macular dystrophies. *Retina* 1985; **5**:172–178.

201. Eagle RC, Lucier AC, Bernardino VB, et al: Retinal pigment abnormalities in fundus flavimaculatus: a light and electron microscopic study. *Ophthalmology* 1980;**87**:1189–1200.

202. Fish G, Grey R, Sehmi KS, et al: The dark choroid in posterior retinal dystrophies. *Br J Ophthalmol* 1981;**65**:359–363.

203. Kandori F: A very rare cause of congenital nonprogressive nightblindness with fleck retina. *Jpn J Ophthalmol* 1959;**13**:384–386.

204. Kandori F, Setogawa T, Tamai A: ERG of new cases of congenital nightblindness with fleck retina. *Jpn J Ophthalmol* 1966;**10**(suppl): 301–316.

205. Kandori F, Tamai A, Kurimoto S, et al: Fleck retina. *Am J Ophthalmol* 1972;**73**: 673–685.

206. Nilsson SEG, Skoog KO: The ERG c-wave in vitelliform macular degeneration (VMD). *Acta Ophthalmol* 1980;**58**:659–666.

207. Rover J, Huttel M, Schaubele G: The DC-ERG: technical problems in recording from patients. *Doc Ophthalmol Proc Ser* 1982;**31**:73–79.

208. Deutman AF: Electro-oculogram in families with vitelliform dystrophy of the fovea: detection of the carrier state. *Arch Ophthalmol* 1969;**81**:305–316.

209. Weingeist TA, Kobrin JL, Watzke RC: Histopathology of Best's macular dystrophy. *Arch Ophthalmol* 1982;**100**:1108–1114.

210. O'Gorman S, Flaherty WA, Fishman GA, et al: Histopathologic findings in Best's vitelliform macular dystrophy. *Arch Ophthalmol* 1988;**106**:1261–1268.

211. Deutman AF, van Blommestein JDA, Henkes HE, et al: Butterfly-shaped pigment dystrophy of the fovea. *Arch Ophthalmol* 1970;**83**:558–569.

212. Marmor MF, Byers B: Pattern dystrophy of the pigment epithelium. *Am J Ophthalmol* 1977;**84**:32–44.

213. Watzke RC, Folk JC, Lang RM: Pattern dystrophy of the retinal pigment epithelium. *Ophthalmology* 1982;**89**:1400–1406.

214. Simell O, Takki K: Raised plasma ornithine and gyrate atrophy of the choroid and retina. *Lancet* 1973;**1**:1031–1033.

215. Weleber RG, Kennaway NG, Brust NRR: Vitamin B_6 in management of gyrate atrophy of choroid and retina. *Lancet* 1978;**2**:1213.

216. Valle D, Walser M, Brusilow SW, et al: Gyrate atrophy of the choroid and retina: amino-acid metabolism and correction of hyperornithinemia with an arginine-deficient diet. *J Clin Invest* 1980;**65**:371–378.

217. Mauthner L: Ein Fall von Choroideremia. *Naturwissenschaften* 1871;**2**:191.

218. Sieving PA, Niffenegger JH, Berson EL: Electroretinographic findings in selected pedigrees with choroideremia. *Am J Ophthalmol* 1986;**101**:361–367.

219. Nussbaum RL, Lewis RA, Lesko JG, et al: Choroideremia is linked to the restriction fragment length polymorphism DXYS1 at Xq 13-21. *Am J Hum Genet* 1985;**37**:473–481.

220. Lewis RA, Nussbaum RL, Ferrell RF: Mapping X-linked ophthalmic diseases: provisional assignment of the locus for choroideremia to Xq 13-24. *Ophthalmology* 1985;**92**: 800–806.

221. François J, de Rouck A: Electrophysiological studies in Groenblad-Strandberg syndrome. *Doc Ophthalmol Proc Ser* 1980;**23**:19–25.

222. Tanino T, Katsumi O, Hirose T: Electrophysiological similarities between two eyes

with X-linked recessive retinoschisis. *Doc Ophthalmol* 1985;60:149–161.

223. Carr RE, Siegel IM: The vitreo-tapeto-retinal degenerations. *Arch Ophthalmol* 1970; **84**:436–445.

224. Lewis RA, Lee GB, Martonyi CL, et al: Familial foveal retinoschisis. *Arch Ophthalmol* 1977;**95**:1190–1196.

225. Shimazaki J, Matsuhashi M: Familial retinoschisis in female patients. *Doc Ophthalmol* 1987;**65**:393–400.

226. Yamaguchi K, Hara S: Autosomal juvenile retinoschisis without foveal retinoschisis. *Br J Ophthalmol* 1989;**73**:470–473.

227. Noble KG, Carr RE, Siegel IM: Familial foveal retinoschisis associated with a rod-cone dystrophy. *Am J Ophthalmol* 1978;**85**:551–557.

228. Fishman GA, Jampol LM, Goldberg MF: Diagnostic features of the Favre-Goldmann syndrome. *Br J Ophthalmol* 1976;**60**:345–353.

229. Hirose T, Lee KY, Schepens C: Wagner's hereditary vitreoretinal degeneration and retinal detachment. *Arch Ophthalmol* 1973;**89**: 176–185.

230. Berson EL, Gouras P, Hoff M: Temporal aspects of the electroretinogram. *Arch Ophthalmol* 1969;**81**:207–214.

231. Cantrill HL, Ramsay RC, Knobloch WH, et al: Electrophysiologic changes in chronic pars planitis. *Am J Ophthalmol* 1981;**91**: 505–512.

232. Ryan SJ, Maumenee AE: Birdshot retinochoroidopathy. *Am J Ophthalmol* 1980;**89**: 31–45.

233. Gass JDM: Vitiliginous chorioretinitis. *Arch Ophthalmol* 1981;**99**:1778–1787.

234. Aaberg TM: Diffuse inflammatory salmon patch choroidopathy syndrome. Presented at International Fluorescein Macula Symposium; Carmel, Calif, October 1979.

235. Fuerst DJ, Tessler HH, Fishman GA, et al: Birdshot retinochoroidopathy. *Arch Ophthalmol* 1984;**102**:214–219.

236. Godel V, Baruch E, Lazar M: Late development of chorioretinal lesions in birdshot retinochoroidopathy. *Ann Ophthalmol* 1989; **21**:49–52.

237. Kaplan HJ, Aaberg TM: Birdshot retinochoroidopathy. *Am J Ophthalmol* 1980;**90**: 773–782.

238. Priem HA, Oosterhuis JA: Birdshot chorioretinopathy: clinical characteristics and evolution. *Br J Ophthalmol* 1988;**72**:646–659.

239. Jampol LM, Sieving PA, Pugh D, et al: Multiple evanescent white dot syndrome, I: clinical findings. *Arch Ophthalmol* 1984;**102**: 671–674.

240. Sieving PA, Fishman GA, Jampol LM, et al: Multiple evanescent white dot syndrome, II: electrophysiology of the photoreceptors during retinal pigment epithelial disease. *Arch Ophthalmol* 1984;**102**:675–679.

241. Peachey NS, Charles HC, Lee CM, et al: Electroretinographic findings in sickle cell retinopathy. *Arch Ophthalmol* 1987;**105**:934–938.

242. Krill AE, Diamond M: The electroretinogram in carotid artery disease. *Arch Ophthalmol* 1962;**68**:42–51.

243. Ulrich WD, Ulrich CH, Kästner R, et al: ERG and EOG in carotid artery occlusion disease. *Doc Ophthalmol Proc Ser* 1980;**23**:49–55.

244. Johnson MA, Marcus S, Elman MJ: ERG sensitivity loss in venous stasis retinopathy. *Invest Ophthalmol Vis Sci* 1987;**28**:319. Abstract.

245. Henkes HE: Electroretinography in circulatory disturbances of the retina: electroretinogram in cases of occlusion of the central retinal artery or one of its branches. *Arch Ophthalmol* 1954;**51**:42–53.

246. Karpe G, Uchermann A: The clinical electroretinogram, VII: the electroretinogram in circulatory disturbances of the retina. *Acta Ophthalmol* 1955;**33**:493–516.

247. Flower RW, Speros P, Kenyon K: Electroretinographic changes and choroidal defects in a case of central retinal artery occlusion. *Am J Ophthalmol* 1977;**83**:451–459.

248. Eichler J, Stave J: Electroretinographic findings in retinal vascular disease. *Doc Ophthalmol Proc Ser* 1980;**23**:57–61.

249. Ponte F: ERG and vascular disturbances. In: François J, ed. *The Clinical Value of Electroretinography*. ISCERG Symposium, Ghent, 1966. Basel: S Karger AG; 1968:300–311.

250. Johnson MA, Marcus S, Elman MJ, et al: Neovascularization in central retinal vein occlusion: electroretinographic findings. *Arch Ophthalmol* 1988;**106**:348–352.

251. Sabates R, Hirose T, McMeel JW: Electroretinography in the prognosis and classification of central retinal vein occlusion. *Arch Ophthalmol* 1983;**101**:232–235.

252. Breton ME, Quinn GE, Keene SS, et al: Electroretinogram parameters at presentation as predictors of rubeosis in central retinal vein occlusion patients. *Ophthalmology* 1989;**96**: 1343–1352.

253. Henkes HE, van der Kam JP: Electroretinographic studies in general arterial hypertension and in arteriosclerosis. *Angiology* 1954; **5**:49–58.

254. Eichler J, Stave J, Bohm J: Oscillatory potentials in hypertensive retinopathy. *Doc Ophthalmol Proc Ser* 1984;**40**:161–165.

255. Muller W, Gaub J, Spittel U, et al: Oscillatory potentials in cases of systemic hypertension. *Doc Ophthalmol Proc Ser* 1984;**40**:167–171.

256. Bernstein HN: Chloroquine ocular toxicity. *Surv Ophthalmol* 1967;**12**:415–442.

257. Finbloom DS, Silver K, Newsome DA, et al: Comparison of hydroxychloroquine and chloroquine use and the development of retinal toxicity. *J Rheumatol* 1985;**12**:692–694.

258. Henkind P, Carr RE, Siegel IM: Early chloroquine retinopathy: clinical and functional findings. *Arch Ophthalmol* 1964;**71**:157–165.

259. Adlakha D, Crews SJ, Shearer ACI, et al: Electrodiagnosis in drug induced disorders of the eye. *Trans Ophthalmol Soc UK* 1967;**87**: 267–284.

260. Sassaman FW, Cassidy JT, Alpern M: Electroretinography in patients with connective tissue diseases treated with hydroxychloroquine. *Am J Ophthalmol* 1970;**70**:515–523.

261. Siddall JR: Ocular toxic changes associated with chlorpromazine and thioridazine. *Can J Ophthalmol* 1966;**1**:190–198.

262. Alkemade PPH: Fenothiazine-retinopathie. *Ned Tijdschr Geneeskd* 1966;**110**:1687.

263. Boet DJ: Toxic effects of phenothiazines on the eye. *Doc Ophthalmol* 1970;**28**:16–69.

264. Busch KT, Busch G, Lehmann W: Zur Frage von Netzhautdegeneration nach Anwendung hoher Phenothiazindosen in der Psychiatrie. *Klin Monatsbl Augenheilkd* 1963; **143**:743–749.

265. Fishman GA: Thioridazine hydrochloride (Mellaril) toxic pigmentary chorioretinopathy. In: Smith JL, ed. *Neuro-ophthalmology Focus*. New York: Masson Publishers; 1982:109–118.

266. Potts AM: Uveal pigment and phenothiazine compounds. *Trans Am Ophthalmol Soc* 1962;**60**:517–552.

267. Davidorf FH: Thioridazine pigmentary retinopathy. *Arch Ophthalmol* 1973;**90**:251–255.

268. Meredith TA, Aaberg TM, Willerson WD: Progressive chorioretinopathy after receiving thioridazine. *Arch Ophthalmol* 1978;**96**: 1172–1176.

269. Appelbaum A: An ophthalmoscopic study of patients under treatment with thioridazine. *Arch Ophthalmol* 1963;**69**:578–580.

270. Goar EL, Fletcher MC: Toxic chorioretinopathy following the use of NP-207. *Am J Ophthalmol* 1957;**44**:603–608.

271. Henkes HE, van Lith GH, Canta LR:

Indomethacin retinopathy. *Am J Ophthalmol* 1972;**73**:846–856.

272. Rothermich NO: Five-year summary of experience with indomethacin. International Symposium on Non-steroid Anti-inflammatory Compounds. Proceedings of the Second Laurentian Rheumatology Conference; Quebec, Canada, 1966. *Excerpta Medica* 1968;**49**:53.

273. Burns CA: Indomethacin, reduced retinal sensitivity and corneal deposits. *Am J Ophthalmol* 1968;**66**:825–835.

274. Cibis GW, Burian HM, Blodi FC: Electroretinogram changes in acute quinine poisoning. *Arch Ophthalmol* 1973;**90**:307–309.

275. Moloney JBM, Hillery M, Fenton M: Two year electrophysiology follow-up in quinine amblyopia: a case report. *Acta Ophthalmol* 1987;**65**:731–734.

276. François J, de Rouck A, Cambie E: Retinal and optic evaluation in quinine poisoning. *Ann Ophthalmol* 1972;**4**:177–185.

277. Bacon P, Spalton DJ, Smith SE: Blindness from quinine toxicity. *Br J Ophthalmol* 1988; **72**:219–224.

278. Potts AM, Proglin J, Farkas I, et al: Studies on the visual toxicity of methanol. *Am J Ophthalmol* 1955;**40**:76–83.

279. Ruedemann AD: The electroretinogram in chronic methyl alcohol poisoning in human beings. *Am J Ophthalmol* 1962;**54**:34–53.

280. Hayreh MS, Hayreh SS, Baumbach GL, et al: Methyl alcohol poisoning, III: ocular toxicity. *Arch Ophthalmol* 1977;**95**:1851–1858.

281. Baumbach GL, Camcilla PA, Martin-Amat G, et al: Methyl alcohol poisoning, IV: alterations of the morphological findings of the retina and optic nerve. *Arch Ophthalmol* 1977; **95**:1859–1865.

282. Peyman GA, May DR, Ericson ES, et al: Intraocular injection of gentamicin: toxic effects and clearance. *Arch Ophthalmol* 1974;**92**:42–47.

283. Palimeris G, Moschos M, Chimonidou E, et al: Intravitreal injection of gentamicin: experimental findings. *Doc Ophthalmol Proc Ser* 1978;**15**:45–52.

284. Mochizuki K, Torisaki M, Kawasaki Y, et al: Retinal toxicity of antibiotics: evaluation by electroretinogram. *Doc Ophthalmol* 1988;**69**: 195–202.

285. Dhanda RP: Electroretinography in night blindness and other vitamin A-deficiencies. *Arch Ophthalmol* 1955;**54**:841–849.

286. Dowling JE: Night blindness, dark adaptation, and the electroretinogram. *Am J Ophthalmol* 1960;**50**:875–889.

287. Genest A: Vitamin A and the electroretinogram in humans. In: François J, ed. *The Clinical Value of Electroretinography*. ISCERG Symposium, Ghent, 1966. Basel: S Karger AG; 1968:250–259.

288. Sandberg MA, Rosen JB, Berson EL: Cone and rod function in vitamin A deficiency with chronic alcoholism and in retinitis pigmentosa. *Am J Ophthalmol* 1977;**84**:658–665.

289. Rushton WA: Rod/cone rivalry in pigment regeneration. *J Physiol* 1968;**198**:219–236.

290. Weleber RG, Denman ST, Hanifin JM, et al: Abnormal retinal function associated with isotretinoin therapy for acne. *Arch Ophthalmol* 1986;**104**:831–837.

291. Kaiser-Kupfer MI, Peck GL, Caruso RC, et al: Abnormal retinal function associated with fenretinide, a synthetic retinoid. *Arch Ophthalmol* 1986;**104**:69–70.

292. Fazio DT, Heckenlively JR, Martin DA, et al: The electroretinogram in advanced open-angle glaucoma. *Doc Ophthalmol* 1986;**63**: 45–54.

293. Fuller DG, Knighton RW, Machemer R: Bright-flash electroretinography for the evaluation of eyes with opaque vitreous. *Am J Ophthalmol* 1975;**80**:214–223.

294. Abrams GW, Knighton RW: Falsely extinguished bright-flash electroretinogram. *Arch Ophthalmol* 1982;**100**:1427–1429.

295. Mandelbaum S, Ober RR, Ogden TE: Nonrecordable electroretinogram in vitreous hemorrhage. *Ophthalmology* 1982;**89**:73–75.

296. Bresnick GH: Electroretinography and color vision in diabetes mellitus. In: Crepaldi G, Cunha-Vaz JG, Fedele D, et al, eds. *Microvascular and Neurological Complications of Diabetes*. Padova: Livina Press; 1987:17–33.

297. Yonemura D, Kawasaki K: Electrophysiological study on activities of neuronal and nonneuronal retinal elements in man with reference to its clinical application. *Jpn J Ophthalmol* 1977;**22**:195–213.

298. Bresnick GH, Palta M: Temporal aspects of the electroretinogram in diabetic retinopathy. *Arch Ophthalmol* 1987;**105**:660–664.

299. Bresnick GH, Palta M: Oscillatory potential amplitudes: relation to severity of diabetic retinopathy. *Arch Ophthalmol* 1987;**105**: 929–933.

300. Bresnick GH, Palta M: Predicting progression to severe proliferative diabetic retinopathy. *Arch Ophthalmol* 1987;**105**:810–814.

301. Bresnick GH, Korth K, Groo A, et al: Electroretinographic oscillatory potentials predict progression of diabetic retinopathy. *Arch Ophthalmol* 1984;**102**:1307–1311.

302. Deneault LG, Kozak WM, Denowski TS: ERG's in streptozotocin-diabetic rats under different insulin regimens. *Doc Ophthalmol Proc Ser* 1980;**23**:67–76.

303. Lawwill T, O'Connor PR: ERG and EOG in diabetics pre and post photocoagulation. *Doc Ophthalmol Proc Ser* 1972;**2**:17–23.

304. Wepman B, Sokol S, Price J: The effects of photocoagulation on the electroretinogram and dark adaptation in diabetic retinopathy. *Doc Ophthalmol Proc Ser* 1977;**13**:139–147.

305. Ogden TE, Callahan F, Riekhof FT: The electroretinogram after peripheral retinal ablation in diabetic retinopathy. *Am J Ophthalmol* 1976;**81**:397–402.

306. Frank RN: Visual fields and electroretinography following extensive photocoagulation. *Arch Ophthalmol* 1975;**93**:591–598.

307. Karpe G, Rendahl I: Clinical electroretinography in detachment of the retina. *Acta Ophthalmol* 1969;**47**:633–641.

308. Rendahl I: The electroretinogram in detachment of the retina. *Arch Ophthalmol* 1957;**57**:566–576.

309. Jacobson JH, Basar D, Carroll J, et al: The electroretinogram as a prognostic aid in retinal detachment. *Arch Ophthalmol* 1958;**59**: 515–520.

310. Schmoger E: Die prognostische Bedeutung des Electroretinogramms bei ablatio retinae. *Klin Monatsbl Augenheilkd* 1957;**131**: 335–343.

311. François J, de Rouck A: L'electroretinographie dans le myopie et les décollements myopigenes de la retine. *Acta Ophthalmol* 1955; **33**:131–155.

312. Wirth A: ERG and endocrine disorders. In: François J, ed. *The Clinical Value of Electroretinography*. ISCERG Symposium, Ghent, 1966. Basel: S Karger AG; 1968:260–266.

313. Pearlman JT, Burian HM: Electroretinographic findings in thyroid dysfunction. *Am J Ophthalmol* 1964;**58**:216–226.

314. Wirth A, Tota G: Electroretinogram and adrenal cortical function. In: Schmoger E, ed. *Advances in Electrophysiology and Pathology of the Visual System*. Sixth ISCERG Symposium, Leipzig, 1968. Stuttgart: George Thieme, Leipzig Verlag; 347–350.

315. Zimmerman TJ, Dawson WW, Fitzgerald CR: Part I: Electroretinographic changes in normal eyes during administration of prednisone. *Ann Ophthalmol* 1973;**5**:757–765.

316. Negi A, Honda Y, Kawano S-I: Why do corticosteroids increase the ERG amplitude? An experimental study in rabbits. *Doc Ophthalmol Proc Ser* 1980;**23**:79–85.

317. Perossini M, Tota G: The effects of aldosterone on the c-wave of the rabbit. *Doc Ophthalmol Proc Ser* 1980;**23**:87–90.

318. Wirth A: Electroretinogram in general medicine. *Ophthalmologica* 1979;**178**:273–288.

319. Burian HM, Burns CA: Electroretinography and dark adaptation in patients with myotonic dystrophy. *Am J Ophthalmol* 1966;**61**:1044–1054.

320. Cavallacci G, Marconcini C, Perossini M: ERG variations in myotonic dystrophy (Steinert's). *Doc Ophthalmol Proc Ser* 1980;**23**:133–136.

321. Stanescu B, Michiels J: Retinal degenerations, electroretinographic aspects in patients with myotonic dystrophy. *Doc Ophthalmol Proc Ser* 1977;**13**:257–262.

322. Stewart HL, Rubin ML: Visual electrophysiology of myotonia congenita. *Doc Ophthalmol Proc Ser* 1972;**2**:5–16.

323. Krill AE, Lee GB: The electroretinogram in albinos and carriers of the ocular albino trait. *Arch Ophthalmol* 1963;**69**:32–38.

324. Bergsma DR, Kaiser-Kupfer M: A new form of albinism. *Am J Ophthalmol* 1974;**77**:837–844.

325. Tomei F, Wirth A: The electroretinogram of albinos. *Vision Res* 1978;**18**:1465–1466.

326. Wack MA, Peachey NS, Fishman GA: Electroretinographic findings in human oculocutaneous albinism. *Ophthalmology* 1989;**96**:1778–1785.

327. Knave B: Electroretinography in eyes with retained intraocular metallic foreign bodies. *Acta Ophthalmol* 1969(suppl 100):3–63.

328. Declercq S, Meredith PC, Rosenthal AR: Experimental siderosis in the rabbit. *Arch Ophthalmol* 1977;**95**:1051–1058.

329. Declercq SS, Phillip CA, Meredith PC: Electrophysiology in experimental siderosis: a follow-up study after removal of intraocular foreign body. *Doc Ophthalmol Proc Ser* 1978;**15**:69–72.

330. Prijot E, Colmant I, Marechol-Courtois C: Electroretinography and myopia. In: François J, ed. *The Clinical Value of Electrophysiology.* ISCERG Symposium, Ghent, 1966. Basel: S Karger AG; 1968:440–443.

331. Hayes KC, Rabin AR, Berson EL: An ultrastructural study of nutritionally induced and reversed retinal degeneration in cats. *Am J Pathol* 1975;**78**:505–515.

332. Schmidt SY, Berson EL, Hayes KC: Retinal degeneration in cats fed casein, I: taurine deficiency. *Invest Ophthalmol Vis Sci* 1976;**15**:47–52.

333. Geggel HS, Heckenlively JR, Martin DA, et al: Human retinal dysfunction and taurine deficiency. *Doc Ophthalmol Proc Ser* 1982;**31**:199–207.

334. Geggel HS, Ament ME, Heckenlively JR, et al: Nutritional requirement for taurine in patients receiving long-term parenteral nutrition. *N Engl J Med* 1985;**312**:142–146.

Supplemental Readings

Arden GB, Bankes JLK: Foveal electroretinogram as a clinical test. *Br J Ophthalmol* 1966;**50**:740.

Armington JC: *The Electroretinogram.* New York: Academic Press; 1974.

Armington JC, Johnson EP, Riggs LA: The scotopic a-wave in the electrical response of the human retina. *J Physiol* 1952;**118**:289–298.

Armington JC, Schwab GJ: Electroretinogram in nyctalopia. *Arch Ophthalmol* 1954;**52**:725–733.

Auerbach E, Burian HM: Studies on the photopic-scotopic relationships in the human electroretinogram. *Am J Ophthalmol* 1955;**40**: 42–60.

Bankes JLK: The foveal electroretinogram. *Trans Ophthalmol Soc UK* 1967;**87**:249–262.

Berson EL: Hereditary retinal diseases: classification with the full-field electroretinogram. In: Lawwill T, ed. *ERG, VER, and Psychophysics*. The Hague: Dr W Junk BV Publishers. *Doc Ophthalmol Proc Ser* 1977;**13**:149–171.

Birch DG: Clinical electroretinography. *Ophthalmol Clin North Am* 1989;**2**:469–497.

Bloome MA, Garcia CA: *Manual of Retinal and Choroidal Dystrophies*. Norwalk, Conn: Appleton-Century-Crofts; 1982.

Burns CA: Ocular effects of indomethacin: slit lamp and electroretinographic (ERG) study. *Invest Ophthalmol* 1966;**5**:325. Abstract.

Carr RE: Primary retinal degenerations. In: Duane TD, ed. *Clinical Ophthalmology*. Philadelphia: JB Lippincott Co; 1976:vol 3, chap 24.

Carr RE, Siegel IM: Electrophysiologic aspects of several retinal diseases. *Am J Ophthalmol* 1964;**58**:95–107.

Carr RE, Siegel IM: *Visual Electrodiagnostic Testing: A Practical Guide for the Clinician*. Baltimore, Md: Williams & Wilkins; 1982.

Chatrian GE, Lettich E, Nelson PL, et al: Computer assisted quantitative electroretinography, I: a standardized method. *Am J EEG Technol* 1980;**20**:57–77.

Deutman AF: *The Hereditary Dystrophies of the Posterior Pole of the Eye*. Springfield, Ill: Charles C Thomas; 1971.

Dowling JE: *The Retina: An Approachable Part of the Brain*. Cambridge, Mass: Belknap Press; 1987:164–186.

Fishman GA: *The Electroretinogram and Electro-oculogram in Retinal and Choroidal Disease*. Rochester, Minn: American Academy of Ophthalmology and Otolaryngology; 1975.

Fishman GA: Hereditary retinal and choroidal diseases: electroretinogram and electro-oculogram findings. In: Peyman GA, Sanders DR, Goldberg MF, eds. *Principles and Practice of Ophthalmology*. Philadelphia: WB Saunders Co; 1980;**2**:857–904.

Franceschetti A, François J, Babel J: *Chorio-retinal Heredodegenerations*. Springfield, Ill: Charles C Thomas; 1974.

François J, de Rouck A: Electroretinography in the diagnosis of congenital blindness. In: François J, ed. *The Clinical Value of Electroretinography*. ISCERG Symposium, Ghent, 1966. Basel: S Karger AG; 1968:451–472.

Fricker SJ, ed: *Electrical Responses of the Visual System*. Boston: Little, Brown & Co; 1969.

Fujina T, Hamasaki DI: The effect of occluding the retinal and choroidal circulation on the electroretinogram of monkeys. *J Physiol* 1965; **180**:837–845.

Galloway NR: Early receptor potential in the human eye. *Br J Ophthalmol* 1967;**51**: 261–264.

Galloway NR: *Ophthalmic Electrodiagnosis*. Philadelphia: WB Saunders Co; 1975.

Goldstein EB, Berson EL: Rod and cone contributions to the human early receptor potential. *Vision Res* 1970;**10**:207–218.

Heckenlively JR: *Retinitis Pigmentosa*. Philadelphia: JB Lippincott Co; 1988.

Henkes HE: Fenothiazine-retinopathie. *Ned Tijdschr Geneeskd* 1966;**110**:789–790.

Jacobson JH: *Clinical Electroretinography*. Springfield, Ill: Charles C Thomas; 1961.

Jaffe MJ, Hommer DW, Caruso RC, et al: Attenuating effects of diazepam on the electroretinogram of normal humans. *Retina* 1989;**9**: 216–225.

Knighton RW, Blankenship GW: Electrophysiological evaluation of eyes with opaque media. *Int Ophthalmol Clin* 1980;**20**:1–19.

Krill AE: Clinical electroretinography. In: Hughes WR, ed. *Year Book of Ophthalmology*. Chicago: Year Book Medical Publishers; 1959–1960:5–27.

Krill AE: The electroretinogram and electro-oculogram: clinical applications. *Invest Ophthalmol* 1970;**9**:600–617.

Krill AE: *Hereditary Retinal and Choroidal Diseases*. New York: Harper & Row; 1972;**1**: 227–295.

Marmor MF, Arden GB, Nilsson SEG, et al: Standard for clinical electroretinography. *Arch Ophthalmol* 1989;**107**:816–819.

Marmor MF, Lurie M: Light-induced electrical responses of the retinal pigment epithelium. In: Zinn KM, Marmor MF, eds. *The Retinal Pigment Epithelium*. Cambridge, Mass: Harvard University Press; 1979:226–244.

Merin S, Auerbach E: The central and peripheral retina in macular degenerations. *Arch Ophthalmol* 1970;**84**:710–718.

Merin S, Auerbach E: Retinitis pigmentosa. *Surv Ophthalmol* 1976;**20**:303–346.

Okun E, Gouras P, Bernstein H: Chloroquine retinopathy. *Arch Ophthalmol* 1963;**69**:59–71.

Pedriel G: Electroretinogram and essential hemeralopia. In: François J, ed. *The Clinical Value of Electroretinography*. ISCERG Symposium, Ghent, 1966. Basel: S Karger AG; 1968:384–392.

Ponte F: Electroretinographic evaluation of the sectoral tapeto-retinal degenerations. In: Nakajima A, ed. *Retinal Degenerations, ERG and Optic Pathways*. Fourth ISCERG Symposium. Tokyo: Marzen Co. *Jpn J Ophthalmol* 1966;**10**(suppl): 282–285.

Potts AM: Electrophysiological measurements. In: Potts AM, ed. *The Assessment of Visual Function*. St Louis: CV Mosby Co; 1972:187–206.

Rendahl I: ERG and retinal detachment. In: François J, ed. *The Clinical Value of Electroretinography*. ISCERG Symposium, Ghent, 1966. Basel: S Karger AG; 1968:435–439.

Ripps H: Night blindness revisited: from man to molecules. The Proctor Lecture. *Invest Ophthalmol Vis Sci* 1982;**23**:588–608.

Steinberg, RH: Monitoring communications between photoreceptors and pigment epithelial cells: effects of "mild" systemic hypoxia. *Invest Ophthalmol Vis Sci* 1987;**28**:1888–1904.

Straub W: ERG in acute and chronic intoxications. In: François J, ed. *The Clinical Value of Electroretinography*. ISCERG Symposium, Ghent, 1966. Basel: S Karger AG; 1968: 273–289.

Straub W: Unilateral retinitis pigmentosa. In: Nakajima A, ed. *Retinal Degenerations, ERG and Optic Pathways*. Fourth ISCERG Symposium. Tokyo: Marzen Co. *Jpn J Ophthalmol* 1966; **10**(suppl):278–281.

Thorner MW, Berk M: Flicker fusion test. *Arch Ophthalmol* 1964;**71**:807–815.

Vinken PJ, Bruyn GW, eds: *Neuroretinal Degenerations: Handbook of Clinical Neurology*. Amsterdam: North-Holland Publishing Co; 1972:vol 13.

Weleber RG, Eisner A: Retinal function and physiological studies. In: Newsome DA, ed. *Retinal Dystrophies and Degenerations*. New York: Raven Press; 1988:21–69.

Young RSL, Price J, Walters JW, et al: Photoreceptor responses of patients with congenital stationary night blindness. *Appl Optics* 1987;**26**: 1390–1394.

Zrenner E, ed: *Special Tests of Visual Function*. Developments in Ophthalmology. Basel: S Karger AG; 1984:vol 9.

CHAPTER 2

The Electro-Oculogram in Retinal Disorders

Gerald Allen Fishman, MD

The electro-oculogram (EOG), a term introduced by Marg[1] in 1951, measures a constantly present standing or resting potential of approximately 6 mV that exists between the cornea and the back of the eye. The current flow is oriented such that voltage at the cornea is positive relative to the posterior pole. The amplitude of this potential is altered by changes in retinal illumination, an observation made initially by Holmgren in 1865. The existence of this potential was discovered in 1849 by Emil DuBois-Reymond, professor of physiology in Berlin. However, it was not until 1954, when Riggs[2] reported abnormal values in a case of a retinal pigmentary degeneration, and 1956, when François et al[3,4] reported abnormal data in various retinal diseases, that this test began to have value as an objective clinical test for retinal function. In 1962, Arden and Fojas[5] noted that the most valuable information was obtained not by the absolute amplitude values of the standing potential (which also varied considerably with levels of illumination, electrode placement, magnitude and velocity of visual excursion, eye prominence, and the corneal-retinal potential itself), but rather by a comparison of the amplitudes under light-adapted and dark-adapted states. Kawasaki et al[6] noted a reduction in the human standing potential after an intravenous infusion of 20% mannitol, while Yonemura et al[7] noted a similar reduction in this potential after an intravenous injection of acetazolamide. These drugs were believed to be useful for the study of patients with diffuse pigmentary retinopathy (mannitol) or others with posterior pole retinal degeneration (acetazolamide). In both the cat[8] and the human,[9] hypoxia of the retina has been shown to produce a rise in the standing potential, while a sudden restoration of oxygen produces an abrupt fall. This finding is likely due to changes in potassium concentration within the subretinal space, presumably as a result of changes in the metabolic rate of the sodium-potassium pump in the photoreceptor cell inner segments.[10]

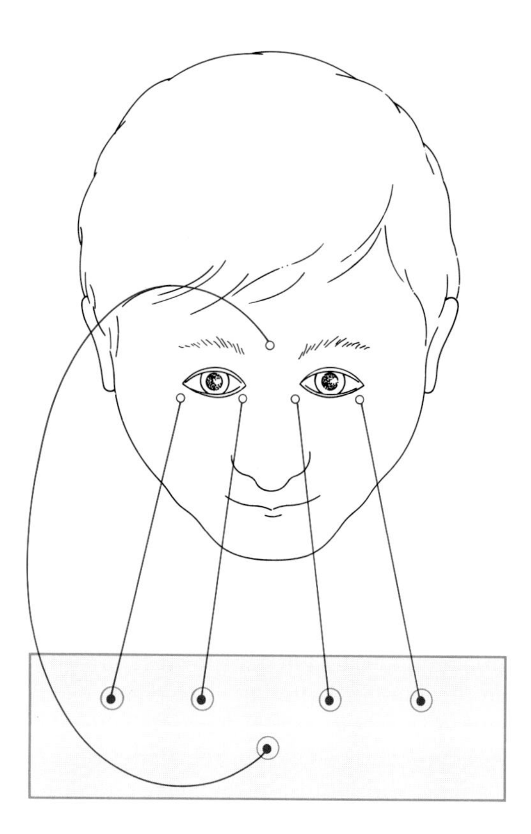

Figure 2-1 *Frontal view of electrode placement for recording EOG responses. Four recording electrode contacts are positioned at the medial and lateral canthi; the ground electrode is placed on the forehead.*

2-1

RECORDING PROCEDURE

The standing potential, which is an overall mass response, is measured clinically by using silver–silver chloride or gold-disk skin electrodes attached with tape near the lateral and medial canthi of both eyes and employing conducting electrode paste or jelly; a similar electrode on the forehead serves as a ground (Figure 2-1). In the electrophysiology laboratory at the University of Illinois at Chicago, the subject sits erect in a lighted room with his head supported by a chin rest and looks into a diffusing sphere. Three small dimly lit red fixation lights are placed in the patient's line of vision so that the center light serves for central fixation and the others allow an excursion of 30° as the patient looks from right to left at the approximate rate of 16–20 sweeps per minute (Figure 2-2). Although the pupils should be dilated to standardize the procedure, a pupil diameter of any value above 3 mm shows little variation in the EOG response. During an initial preadaptation period of approximately 5 minutes, the subject is exposed to continuous room light or bowl background levels while baseline recordings are obtained. The skin electrodes are connected to either a polygraph recorder or an oscilloscope, and ocular movement is recorded as an electrical potential difference between the electrodes at the medial and lateral canthi (hence the term electro-oculogram), which in turn is used to monitor the effects of changes in illumination on the standing potential. After the standardized period of light adaptation, all lights are turned off

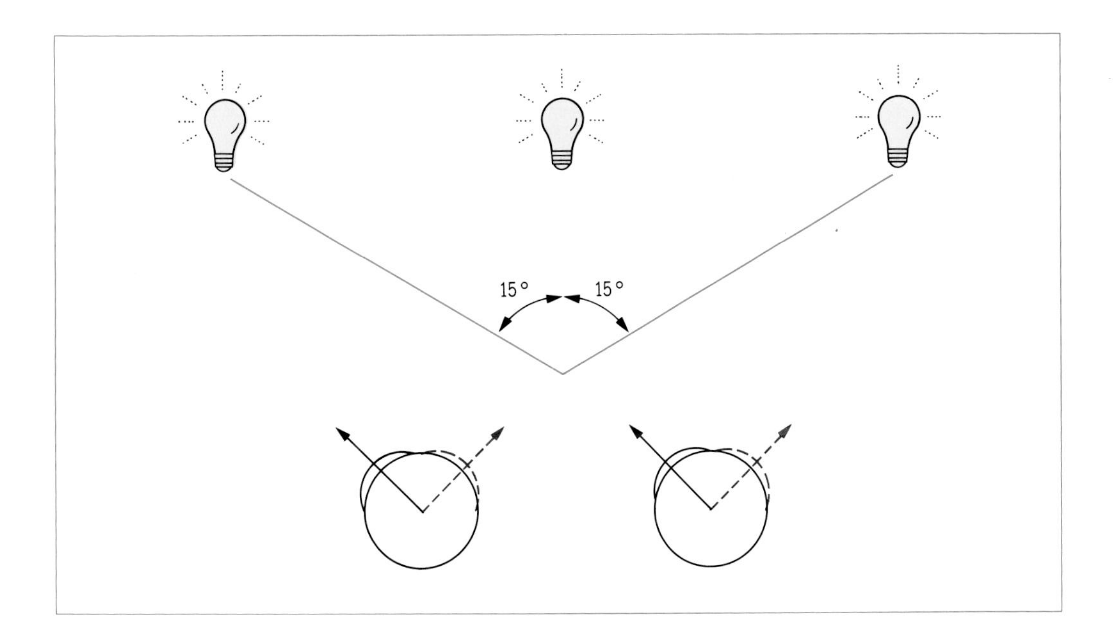

Figure 2-2 *Schematic drawing of fixation lights and ocular excursions used during recording of EOG potentials.*

and responses from saccadic eye movements are recorded for 15 minutes under a dark-adapted condition. After this allotted time, adapting lights from within a diffusing sphere or other light-adapting source are turned on and responses are recorded for another 15 minutes under light-adapted conditions. Arden and Kelsey[11] indicated that at least 2500 trolands should be used to obtain optimal light-adapted responses. With a dilated pupil, this corresponds to an adapting luminance of at least 50 cd/m^2; with a natural pupil, an adaptation luminance of 500 cd/m^2 is required.[12] Weleber and Eisner[13] suggested that with dilated pupils 20 footlamberts (approximately 70 cd/m^2) of ganzfeld illumination is both adequate for obtaining a light peak and well tolerated by patients. Sample recordings are taken at approximately 1-minute intervals under

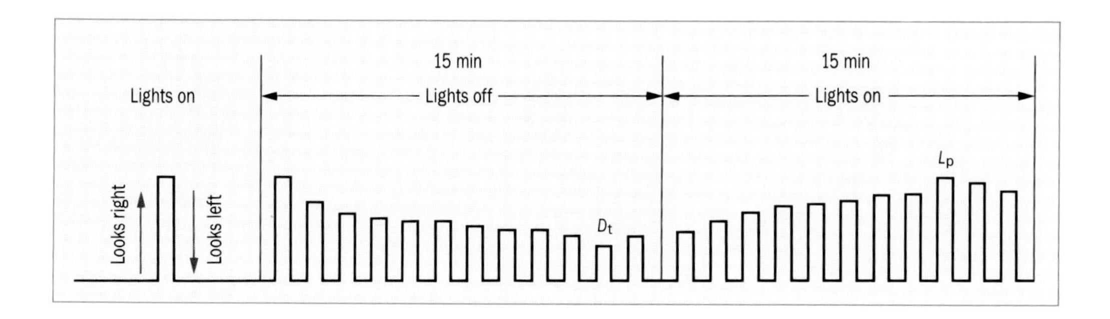

Figure 2-3 *Schematic depiction of EOG potentials recorded as subject looks from right to left between fixation lights under dark- and light-adapted conditions.* L_p *= light peak;* D_t *= dark trough.*

both light-adapted and dark-adapted conditions. Figure 2-3 outlines this response sequence.

Although the amplitudes of the saccadic ocular movements between the fixation points remain constant, the recorded responses progressively decrease in the dark, reaching a trough in approximately 8–12 minutes. With light adaptation, there is a progressive rise in amplitude, reaching a peak in approximately 6–9 minutes. The ratio of the light peak (L_p) to dark trough (D_t) is then determined to evaluate the normalcy of the response. Although there is a definite variation, requiring each laboratory to establish its own set of normal values, most investigators find that normal patients have ratios of 1.80 or greater. Values of 1.65–1.80 are probably marginally subnormal, while a ratio of less than 1.65 is considered distinctly subnormal. The reproducibility of ratios in patients can vary, although some investigators maintain that a difference of more than 0.20 in repeated testing of normal patients is unusual. Thus, there is an acceptable test-retest variability of up to 10% of the light-peak to dark-trough ratio.[14] Jones et al[15] noted that males had, on average, a 0.21 lower EOG ratio than did females when they compared 25 normal subjects of each sex who were of a similar age range and had a similar range of refractive errors. These authors also reported that 95% of eyes retested from 20 subjects showed an intersession variability of 0.6 or less for the EOG ratio.

2-2

COMPONENTS AND ORIGINS OF THE EOG

From the preceding discussion and illustrations, it is apparent that the EOG response has two components: one light-insensitive, the other light-sensitive. The light-insensitive component depends on the integrity of the pigment epithelium as well as some extraretinal sources such as the cornea, lens, and ciliary body. This light-insensitive component, accounting for the dark trough, is not influenced by previous retinal illumination and thus is independent of the functional status of the retinal photoreceptors. The light-sensitive component, or slow light rise of the EOG, appears to be generated by a depolarization of the basal membrane of the retinal pigment epithelium, which is not directly related to any change in retinal potassium concentration.[16] This depolarization of the basal membrane results in an increase in the electrical potential across retinal pigment epithelial cells (transepithelial potential). Intact photoreceptor cells are, however, also necessary to generate this response. Further, this light rise response requires physical contact to be maintained between the photoreceptors and the pigment epithelium and thus it is not retained in the presence of a detached retina. A contribution from inner nuclear layer cells is suggested by the finding that the light rise can be reduced after occlusion of the central retinal artery, which theoretically should have no direct effect on potentials arising from the retinal pigment epithelium or photoreceptors.[17] The light rise, and thus the EOG light-peak to dark-trough ratio, do not correlate with the degree, pattern, or even presence of melanin within retinal pigment epithelial cells. In the arterially perfused cat eye, an increase in extracellular calcium was demonstrated to cause a decrease in the light peak by 85–90%.[18]

Of note, hyperpolarization, rather than depolarization, of the basal membrane of the retinal pigment epithelium has been measured by EOG recordings as a slight, fast, and transient decline in voltage of the transepithelial potential after light onset, which contrasts with the light rise or slow oscillation seen after light onset. This fast oscillation does relate to changes in potassium concentration within the retina.[19] The amplitude of the light-sensitive EOG component increases with increasing levels of retinal illumination within the range of 1–1000 foot-lamberts. At the higher levels of illumination, the ratio manifests more variability. The EOG light rise is not mediated exclusively by the rod or cone systems, since normal values are generally obtained in congenital achromats and in patients with congenital stationary night blindness. Further clinical and basic science investigations are necessary to define more precisely the entire physiologic basis of the various EOG components.

Kawasaki et al[6] noted that the amplitude of the EOG, after stabilization in the dark, can be reduced by an average of 43% subsequent to the intravenous injection of a hypertonic solution of 20% mannitol. This finding, referred to as the hyperosmolarity-induced response, is directly attributable to alterations in function at the level of the retinal pigment epithelium and thus could serve as an objective, quantitative test for selectively determining the function of pigment epithelial cells. Similarly, the function of these cells can be monitored by measuring the reduction in EOG amplitude following the intravenous injection of 500 mg of acetazolamide (*Diamox*).[7] The acetazolamide response reflects primarily the function of retinal pigment epithelial cells within the posterior pole of the fundus and thus can be abnormal in at least some patients with retinal pigment epithelial changes associated with macular degeneration while remaining normal in patients with retinitis pigmentosa. The acetazolamide-induced response and the hyperosmolarity-induced response, although both measuring functional alterations at the level of the retinal pigment epithelium, appear to measure different aspects of cell function.

2-3

THE EOG IN RETINAL DISORDERS

Although the various components of the EOG are not precisely understood, their variability in various disease states provides both a better understanding of, and a way to measure, retinal malfunction. The following section describes the deviations from normal in the EOG caused by various retinal disorders. Emphasis is on the EOG findings in diseases that primarily affect the retinal pigment epithelium. Only brief mention will be made of the various photoreceptor, choroidal, inflammatory, circulatory, toxic, optic nerve, and miscellaneous disorders that might affect the EOG.

2-4

PRESUMED PRIMARY DYSTROPHIES OF THE PIGMENT EPITHELIUM

2-4-1 Stargardt's Macular Dystrophy (Fundus Flavimaculatus)

The clinical features seen in patients with Stargardt's macular dystrophy were described in Section 1-12-1 of Chapter 1. Although initial investigators reported abnormal EOG ratios in the majority of patients with fundus flavimaculatus,[20] subsequent studies showed that abnormal ratios were consistently found primarily in advanced stages of the disease.[21] Further, the recording procedure (ganzfeld bowl versus a flat-surfaced view box) was found to influence whether normal or abnormal EOG responses were obtained in these patients.[22]

2-4-2 Dominantly Inherited Drusen of Bruch's Membrane

Although initial studies reported reduced EOG light-peak to dark-trough ratios in patients with dominantly inherited familial drusen of Bruch's membrane,[23] subsequent investigations showed that the majority have EOG ratios within the normal range, particularly if a ganzfeld bowl, rather than a flat-surfaced view box, is used for the recordings.[24] Sunness and Massof[25] also reported normal EOG findings in patients with drusen as part of age-related macular degeneration when a focal EOG procedure was used.

2-4-3 Best's (Vitelliform) Macular Dystrophy

Best's dystrophy manifests an appreciably reduced EOG ratio not only at an early stage in patients with evident disease[26,27] but also in those who have inherited the gene yet show no clinically apparent fundus changes.[28] In instances when only one eye manifests a clinically apparent retinal lesion, both eyes will show reduced EOG ratios.[26,27] Further, normal EOG ratios help distinguish patients with other disorders, such as adult-onset foveomacular pigment epithelial dystrophy[29] and pseudo-Best's dystrophy,[30] that manifest vitelliform lesions phenotypically similar to those seen in Best's dystrophy.

2-4-4 Butterfly Dystrophy

As one of the pattern dystrophies, butterfly dystrophy is presumed to involve a primary disease of the pigment epithelium. It is therefore not surprising that (generally small) reductions in EOG light-peak to dark-trough ratios have been reported. Nevertheless, this finding is not universal, as a number of patients have normal EOG ratios. Thus, the EOG is either not impaired or often not sufficiently impaired to be of diagnostic value in this disorder. This entity was described in greater detail in Section 1-12-4 of Chapter 1.

2-5

PHOTORECEPTOR DYSTROPHIES OR DYSFUNCTIONS

2-5-1 Rod-Cone

The EOG is often abnormal in patients with retinitis pigmentosa. In early stages, variable results are obtained when attempting to evaluate the diagnostic sensitivity of the EOG over the ERG amplitudes. Although in later stages the two tests generally parallel each other in severity of impairment, Gouras and Carr[31] noted that abnormal EOG ratios paralleled abnormal cone ERG functions and both were preceded by abnormal rod ERG functions in patients with early retinitis pigmentosa. In patients with retinitis pigmentosa and appreciable reductions in ERG rod and cone function, measurement of EOG ratios provides little if any important additional diagnostic information.

2-5-2 Cone and Cone-Rod

Abnormal EOGs have been noted in some patients with acquired cone and cone-rod dystrophies.[32,33] Congenital achromats tend to have normal EOG values. Similarly, patients with progressive diffuse cone dystrophy tend to show normal EOG light-peak to dark-trough ratios, while those with diffuse rod as well as cone disease show a reduction in EOG ratios.

2-6

STATIONARY NIGHT-BLINDING DISORDERS

2-6-1 Congenital Stationary Night Blindness With Myopia and Normal-Appearing Retina

The EOG appears normal in autosomal recessive and X-linked recessive congenital stationary nyctalopia with the Schubert-Bornschein type of ERG waveform.[34] A reduced light rise occurs in the autosomal dominant form with the Riggs type of ERG waveform.

2-6-2 Oguchi's Disease

Although extensive reports are not available, EOG findings in patients with Oguchi's disease appear normal.[35]

2-6-3 Fundus Albipunctatus

In patients with fundus albipunctatus, the EOG will not show a notable light rise if the procedure is performed in the usual manner, with a period of dark adaptation of approximately 15 minutes. However, a normal light rise and EOG ratio will be found if a dark-adaptation period of several hours is obtained prior to light exposure.

2-7

CHOROIDAL DYSTROPHIES

Choroideremia, gyrate atrophy, and diffuse choroidal atrophy patients show abnormal EOG values. Generally, the abnormal values are noted early in these diseases and often parallel the extent of diminished ERG amplitudes.

2-8

INFLAMMATORY CONDITIONS

As with the ERG, local inflammatory disease does not affect the EOG. In the presence of diffuse, generally chronic chorioretinal inflammation, the EOG light-peak to dark-trough ratio is subnormal, its degree of abnormality generally correlating with the extent of clinically apparent disease and its noted effect on the ERG.

2-9

CIRCULATORY DEFICIENCIES

Occlusive disease of the central retinal artery and its effect on the EOG light rise have been mentioned in this chapter in Section 2-2. Circulatory deficiencies within the carotid, ophthalmic, and choroidal vessels would be expected to affect the EOG ratio, although Ashworth[36] reported the EOG to be normal in 3 of 4 eyes on the side of a carotid occlusion. In 5 eyes with hypertensive retinopathy, also evaluated by Ashworth,[36] the EOG was clearly abnormal in 3. Normal or abnormal findings are likely to depend on the degree of induced retinal hypoxia.

2-10

TOXIC CONDITIONS

2-10-1 Chloroquine and Hydroxychloroquine

Abnormal findings in EOG as well as ERG recordings, such as a reduction in the light-peak to dark-trough ratio and reduced amplitudes, respectively, have been reported in patients receiving antimalarial drugs such as chloroquine or hydroxychloroquine. These changes tend to occur after prolonged administration of these drugs and when funduscopic evidence of foveal and often also peripheral retinal pigmentary changes are already in evidence. Thus, these electrophysiologic tests are not likely to ascertain patients who are developing retinopathy at a sufficiently early stage to warrant their use in monitoring patients for signs of early retinal toxicity. As noted in the discussion on the ERG and toxic retinopathy (*Section 1-18 in Chapter 1*), static perimetry testing within the central 10° would provide a more reliable means of monitoring such patients. Pinckers and Broekhuyse[37] noted that subnormal EOG ratios were recorded in only 37% of patients with known retinal toxicity and a bull's-eye maculopathy as the consequence of treatment with chloroquine or hydroxychloroquine. Gouras and Gunkel[38] also noted that the EOG was not a sensitive indicator of mild degrees of chloroquine retinopathy, particularly in patients no longer receiving chloroquine therapy. Pinckers and Broekhuyse[37] also suggested that an underlying systemic disorder, such as rheumatoid arthritis, may in itself contribute to an abnormal EOG ratio, independent of the use of antimalarial drugs. They noted the EOG to be subnormal in 20% of such patients. Therefore, the interindividual and intraindividual variability of the EOG, as well as its questionable diagnostic sensitivity, precludes its value as a means of predictably diagnosing early functional impairment or in monitoring for progressive retinal deterioration those patients receiving these antimalarial drugs.

2-10-2 Siderosis Bulbi

Patients with siderosis bulbi show EOG abnormalities. Since the number of cases studied has not been extensive, it is unclear how early in disease progression this change might be observed. Some investigators maintain that the EOG is a sensitive way of monitoring the early changes of siderosis.

2-11

OPTIC NERVE DISEASE

As might be expected, disease of primarily the optic nerve generally does not affect the EOG light-peak to dark-trough ratio.

2-12

MISCELLANEOUS CONDITIONS

2-12-1 Retinal Detachment

As with the ERG, the degree of retinal detachment is reflected in a progressively abnormal EOG ratio. This abnormal ratio indicates the necessity for maintaining the physical contiguity of the rod and cone outer segments and the retinal pigment epithelium for proper generation of EOG responses.

2-12-2 Diabetes

Henkes and Houtsmuller[39] reported that an abnormal EOG may occur in patients with diabetes before changes in the ERG amplitude or eye grounds are apparent. The EOG allegedly deteriorates with the duration of diabetes and, consequently, with the severity of the retinopathy. This area has not, however, been extensively investigated. Kawasaki et al[6] reported an abnormality in the hyperosmolarity-induced response to intravenous mannitol in patients with diabetic retinopathy.

2-12-3 Myopia

Since histologic changes occur in malignant myopia in both the retinal pigment epithelium and the choroid, an abnormal EOG ratio is anticipated and, in fact, found in some patients with the typical fundus changes associated with progressive high myopia. However, Thaler et al[40] noted a linear decrease of the light peak with increasing axial length of the eye, even in the absence of clinically apparent degenerative changes of the retina associated with myopia.

2-12-4 Choroidal Malignant Melanoma

A reduction in the EOG light-peak to dark-trough ratio in patients with a choroidal melanoma was first reported by Ponte and Lauricella.[41] Subsequent investigators noted similar findings,[42-44] and two studies have suggested that EOG measurements may be useful in differentiating between a choroidal nevus and a malignant melanoma.[42,44] Further studies are necessary to determine the physiologic basis for these observations.

2-13

SUMMARY

The EOG has been applied widely to the study of various retinal disorders. Its use in disorders presumed to affect primarily the retinal pigment epithelium, such as Best's macular dystrophy and the various pattern dystrophies, appears most appropriate. Its use in the various stationary and progressive night-blinding disorders seems questionable since the ERG either provides more diagnostically definitive information or allows the investigator to often predict what the EOG findings are likely to be.

REFERENCES

1. Marg E: Development of electro-oculography. *Arch Ophthalmol* 1951;**45**:169–185.

2. Riggs LA: Electroretinography in cases of night blindness. *Am J Ophthalmol* 1954;**38**: 70–78.

3. François J, Verriest G, de Rouck A: Electro-oculography as a functional test in pathological conditions of the fundus, I: first results. *Br J Ophthalmol* 1956;**40**:108–112.

4. François J, Verriest G, de Rouck A: Electro-oculography in fundus pathology. *Br J Ophthalmol* 1956;**40**:305–312.

5. Arden GB, Fojas MR: Electrophysiological abnormalities in pigmentary degenerations of the retina. *Arch Ophthalmol* 1962;**68**:369–389.

6. Kawasaki K, Yamamoto S, Yonemura D: Electrophysiological approach to clinical test for the retinal pigment epithelium. *Acta Soc Ophthalmol Jpn* 1977;**81**:1303–1312.

7. Yonemura D, Kawasaki K, Tanabe J, et al: Susceptibility of the standing potential of the eye to acetazolamide and its clinical applications. *Folia Ophthalmol Japonica* 1978;**29**: 408–416.

8. Linsenmeier RA, Mines AH, Steinberg RH: Effects of hypoxia and hypercapnia on the light peak and electroretinogram of the cat. *Invest Ophthalmol Vis Sci* 1983;**24**:37–46.

9. Marmor MF, Donovan WJ, Gaba DM: Effects of hypoxia and hyperoxia on the human standing potential. *Doc Ophthalmol* 1985;**60**: 347–352.

10. Linsenmeier RA, Steinberg RH: Mild hypoxia alters K^+ homeostasis and pigment epithelial cell membrane responses in the cat retina. *Invest Ophthalmol Vis Sci* 1984; **25**(suppl):289.

11. Arden GB, Kelsey JH: Some observations on the relationship between the standing potential of the human eye and the bleaching and regeneration of visual purple. *J Physiol (Lond)* 1962;**161**:205–226.

12. De Groot SG, Gebhard JW: Pupil size as determined by adapting luminance. *J Opt Soc Am* 1952;**42**:492–495.

13. Weleber RG, Eisner A: Retinal function and physiological studies. In: Newsome DA, ed. *Retinal Dystrophies and Degenerations.* New York: Raven Press; 1988:21–69.

14. Kolder HE, Hochgesand P: Empirical model of the electro-oculogram. *Doc Ophthalmol* 1973;**34**:229–241.

15. Jones RM, Stevens TS, Gould S: Normal EOG values of young subjects. *Doc Ophthalmol Proc Ser* 1977;**13**:93–97.

16. Steinberg RH, Griff ER, Linsenmeier RA: The cellular origin of the light peak. *Doc Ophthalmol Proc Ser* 1983;**37**:1–11.

17. Thaler A, Heilig P, Lessel MR: Ischemic retinopathy: reduced light-peak and dark trough amplitudes in electrooculography. In: Lawwill T, ed. *ERG, VER, and Psychophysics.* The Hague: Dr W Junk BV Publishers. *Doc Ophthalmol Proc Ser* 1977;**13**:119–121.

18. Hofmann H, Niemeyer G: Calcium blocks selectively the EOG light peak. *Doc Ophthalmol* 1985;**60**:361–368.

19. Griff ER, Linsenmeier RA, Steinberg RH: The cellular origin of the fast oscillation. *Doc Ophthalmol Proc Ser* 1983;**37**:13–20.

20. Klien BA, Krill AE: Fundus flavimaculatus: clinical, functional and histopathologic observations. *Am J Ophthalmol* 1967;**64**:3–23.

21. Fishman GA: Fundus flavimaculatus: a clinical classification. *Arch Ophthalmol* 1976;**94**:2061–2067.

22. Fishman GA, Young RSL, Schall SP, et al: Electro-oculogram testing in fundus flavimaculatus. *Arch Ophthalmol* 1979;**97**:1896–1898.

23. Krill AE, Klien BA: Flecked retina syndrome. *Arch Ophthalmol* 1965;**74**:496–508.

24. Fishman GA, Carrasco C, Fishman M: The electro-oculogram in diffuse (familial) drusen. *Arch Ophthalmol* 1976;**94**:231–233.

25. Sunness JS, Massof, RW: Focal electro-oculogram in age-related macular degeneration. *Am J Optom Physiol Optics* 1986;**63**:7–11.

26. François J, de Rouck A, Fernandez-Sasso D: Electro-oculography in vitelliform degeneration of the macula. *Arch Ophthalmol* 1967;**77**: 726–733.

27. François J, de Rouck A, Fernandez-Sasso D: Electroretinography and electro-oculography in diseases of the posterior pole of the eye. *Adv Ophthalmol* 1969;**21**:132–163.

28. Deutman AF: Electro-oculogram in families with vitelliform dystrophy of the fovea: detection of the carrier state. *Arch Ophthalmol* 1969;**81**:305–316.

29. Vine AK, Schatz H: Adult-onset foveomacular pigment epithelial dystrophy. *Am J Ophthalmol* 1980;**89**:680–691.

30. Fishman GA, Trimble S, Rabb MF, et al: Pseudovitelliform macular degeneration. *Arch Ophthalmol* 1977;**95**:73–76.

31. Gouras P, Carr RE: Electrophysiological studies in early retinitis pigmentosa. *Arch Ophthalmol* 1964;**72**:104–110.

32. Krill AE, Deutman AF, Fishman M: The cone degenerations. *Doc Ophthalmol* 1973;**35**: 1–80.

33. Krill AE: The electroretinographic and electro-oculographic findings in patients with macular lesions. *Trans Am Acad Ophthalmol Otolaryngol* 1966;**70**:1063–1083.

34. Volker-Dieben HJ, van Lith GHM, Went LN, et al: Electro-ophthalmology of a family with X-chromosomal recessive nyctalopia and myopia. In: Dodt E, Pearlman T, eds. XIth ISCERG Symposium. The Hague: Dr W Junk BV Publishers. *Doc Ophthalmol Proc Ser* 1974; **4**:169–177.

35. Carr RE, Ripps H: Rhodopsin kinetics and rod adaptation in Oguchi's disease. *Invest Ophthalmol* 1967;**6**:426–436.

36. Ashworth B: The electro-oculogram in disorders of the retinal circulation. *Am J Ophthalmol* 1966;**61**:505–508.

37. Pinckers A, Broekhuyse RM: The EOG in rheumatoid arthritis. *Acta Ophthalmol* 1983;**61**: 831–837.

38. Gouras P, Gunkel RD: The EOG in chloroquine and other retinopathies. *Arch Ophthalmol* 1963;**70**:629–639.

39. Henkes HE, Houtsmuller AJ: Fundus diabeticus: an evaluation of the preretinopathic stage. *Am J Ophthalmol* 1965;**60**:662–670.

40. Thaler A, Heilig P, Scheiber V, et al: The influence of myopia on the EOG. *Doc Ophthalmol Proc Ser* 1974;**10**:325–328.

41. Ponte F, Lauricella M: On the lack of correlation between ERG and EOG alterations in malignant melanoma of the choroid. In: Lawwill T, ed. *ERG, VER, and Psychophysics*. The Hague: Dr W Junk BV Publishers. *Doc Ophthalmol Proc Ser* 1977;**13**:87–92.

42. Staman JA, Fitzgerald CR, Dawson WW, et al: The EOG and choroidal malignant melanomas. *Invest Ophthalmol Vis Sci* 1979; **18**(suppl):121.

43. Jones RM, Klein R, De Venecia G, et al: Abnormal electro-oculograms from eyes with a malignant melanoma of the choroid. *Invest Ophthalmol Vis Sci* 1981;**20**:276–279.

44. Markoff J, Shaken E, Shields J, et al: The electro-oculogram in presumed choroidal melanomas and nevi. In: Niemeyer G, Huber CH, eds. *Techniques in Clinical Electrophysiology of Vision*. The Hague: Dr W Junk BV Publishers. *Doc Ophthalmol Proc Ser* 1982;**31**:105.

Supplemental Readings

Arden GB, Barrada A, Kelsey JH: New clinical test of retinal function based upon the standing potential of the eye. *Br J Ophthalmol* 1962;**46**: 449–467.

Carr RE, Siegel IM: *Visual Electrodiagnostic Testing: A Practical Guide for the Clinician*. Baltimore, Md: Williams & Wilkins; 1982.

Fishman GA: *The Electroretinogram and Electro-oculogram in Retinal and Choroidal Disease.* Rochester, Minn: American Academy of Ophthalmology and Otolaryngology; 1975.

Fishman GA: Hereditary retinal and choroidal diseases: electroretinogram and electro-oculogram findings. In: Peyman GA, Sanders DR, Goldberg MF, eds. *Principles and Practice of Ophthalmology*. Philadelphia: WB Saunders Co; 1980;**2**:857–904.

Fricker SJ, ed: *Electrical Responses of the Visual System*. Boston: Little, Brown & Co; 1969.

Galloway NR: *Ophthalmic Electrodiagnosis*. Philadelphia: WB Saunders Co; 1975.

Gouras P: Relationships of the electrooculogram to the electroretinogram. In: François J, ed. *The Clinical Value of Electroretinography.* ISCERG Symposium, Ghent, 1966. Basel: S Karger AG; 1968:66–73.

Kelsey JH: Variations in the normal electrooculogram. *Br J Ophthalmol* 1967;**51**:44–49.

Krill AE: The electroretinogram and electro-oculogram: clinical applications. *Invest Ophthalmol* 1970;**9**:600–617.

Krill AE: *Hereditary Retinal and Choroidal Diseases*. New York: Harper & Row; 1972;**1**:227–295.

Marmor MF, Lurie M: Light-induced electrical responses of the retinal pigment epithelium. In: Zinn KM, Marmor MF, eds. *The Retinal Pigment Epithelium*. Cambridge, Mass: Harvard University Press; 1979:226–244.

Potts AM: Electrophysiological measurements. In: Potts AM, ed. *The Assessment of Visual Function*. St Louis: CV Mosby Co; 1972:187–206.

Reeser F, Weinstein GW, Feiock KB, et al: Electro-oculography as a test of retinal function. *Am J Ophthalmol* 1970;**70**:505–514.

Steinberg, RH: Monitoring communications between photoreceptors and pigment epithelial cells: effects of "mild" systemic hypoxia. *Invest Ophthalmol Vis Sci* 1987;**28**:1888–1904.

Weleber RG, Eisner A: Retinal function and physiological studies. In: Newsome DA, ed. *Retinal Dystrophies and Degenerations*. New York: Raven Press; 1988:21–69.

Zrenner E, ed: *Special Tests of Visual Function.* Developments in Ophthalmology. Basel: S Karger AG; 1984:vol 9.

CHAPTER 3

The Visually Evoked Cortical Potential in Optic Nerve and Visual Pathway Disorders

Samuel Sokol, PhD

This chapter reviews the fundamentals of visually evoked cortical potential (VECP) recording, illustrates its use as a diagnostic procedure, and provides practical suggestions for establishing a clinical VECP laboratory. Before specific clinical applications are described, a brief overview of VECP fundamentals is presented.

3-1

ORIGINS OF THE VECP

The VECP is a gross electrical signal generated at the visual cortex in response to visual stimulation. Compared to the electroencephalogram (EEG), which represents ongoing activity of the entire cortex, the VECP is specific to occipital responses. Further, the VECP amplitude is smaller (1–40 μV) and more responsive to stimulus changes than the EEG amplitude, which can be as large as 100 μV. Because of this, computer averaging is necessary.

The VECP reflects the electrical activity of a subject's central visual field. Since projections from the central retina are sent to the surface of the occipital lobe and those from the peripheral retina lie deep within the calcarine fissure, an electrode attached to the back of the scalp more effectively records activity from cortical cells receiving central retinal input. A second reason that the VECP reflects activity of the central visual field is that the foveal projection is magnified at the cortex.[1,2] The central part of the visual field thus has a much larger cortical representation than do the peripheral regions. This representation is expressed as the cortical magnification factor (M), which indicates the linear extent of cortex in millimeters corresponding to one degree of visual angle at various eccentricities from the fovea.[2] For example, at the fovea M = 5.6 mm/degree, while 10° from the fovea M = 1.5 mm/degree.[1] Figure 3-1 shows

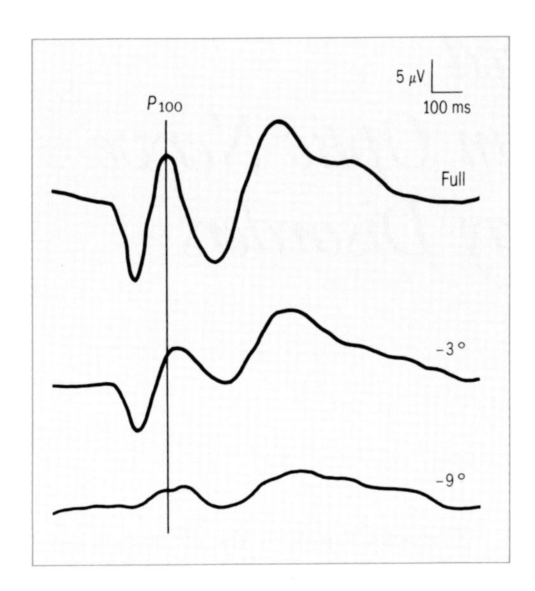

Figure 3-1 *Transient pattern VECPs recorded from a visually normal adult. The top trace is a typical response to full-field stimulation (12-minute checks, 12° × 14° field, two alternations per second). The P_{100} component of the response is indicated. The middle trace (−3°) shows the response when the central 3° of the stimulus is blocked out. Note both the reduced amplitude and the increased latency when a central scotoma is present. The bottom trace (−9°) shows the response when the central 9° is blocked out. In this and all subsequent figures, positive is up.* *Reprinted by permission from Sokol S: Visual evoked potentials. In: Aminoff MJ, ed.* Electrodiagnosis in Clinical Neurology. *2nd ed. New York: Churchill Livingstone; 1986:441–466.*

how the introduction of a central scotoma reduces VECP amplitude.

Within the central region, the fovea and the parafovea can be stimulated differentially by varying check size. Bodis-Wollner et al[3] pointed out that checks of 10–15 minutes stimulate the fovea optimally, while larger checks, such as 50 minutes, stimulate parafoveal regions. A clinical protocol with two check sizes will therefore have a greater diagnostic yield than will a protocol with only one check size.

3-2

VECP STIMULI

The VECP is typically recorded in response to flash or pattern stimuli. Flash stimuli are produced most frequently with xenon-arc photostimulators with which the luminance, stimulus rate, and color of the flash can be varied. Pattern stimuli are usually generated on an oscilloscope or a video monitor. The field size, pattern size, contrast, retinal location, and rate of presentation of pattern stimuli can be varied. Checkerboards are used most often in clinical settings, although square or sine wave gratings can also be used.[3]

VECP pattern stimuli are either phase-reversed (also called pattern reversal, contrast reversal, or counterphase modulation) or flashed on and off. In either case, the overall mean luminance of the stimulus is constant, minimizing luminance contamination of the VECP. For example, when a phase-reversal stimulus is viewed, checks are visible at all times—with half the checks increasing in luminance while the other half decrease. The net result is a constant mean luminance.[4] If the modulation of the checks is asymmetric—ie, one cycle of checks is brighter than the other cycle—the pattern VECP waveform will be contaminated with luminance responses. In the on-off mode, the pattern appears for a discrete length of time (eg, 500 milliseconds) and then is replaced by an unpatterned, diffuse field of the same average luminance as the pattern stimulus.[5]

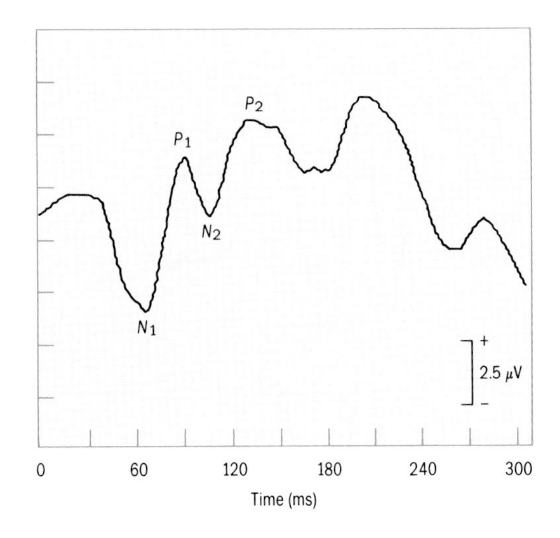

Figure 3-2 *Flash-elicited VECP obtained with a xenon-arc photostimulator set at 1 flash per second. The major components of the waveform are indicated.*

3-3

VECP PARAMETERS

The VECP parameter that is measured depends on the questions being asked and the type of stimulus used to generate the VECP. If the issue is a patient's acuity, amplitude is measured. If a covert lesion of the visual pathway is at issue, then latency (or phase) is measured.

Amplitude is the size, in microvolts, of a particular component. Transient VECP amplitude components that are commonly measured include N_1–P_1 and P_1–N_2. (See Regan[6] for a discussion of the pitfalls of component analysis.) Latency is the time, in milliseconds, from stimulus onset to the peak of a component. The first positive component (P_1), which occurs at about 100 milliseconds, is usually measured in clinical testing. An example of a transient flash VECP obtained with a xenon flash photostimulator is shown in Figure 3-2. Transient pattern VECPs obtained with a video monitor are shown in Figure 3-3.

Absolute latency is more reliable than absolute amplitude. A normal subject shows a latency variability of 2–5% within and between recording sessions, while

A

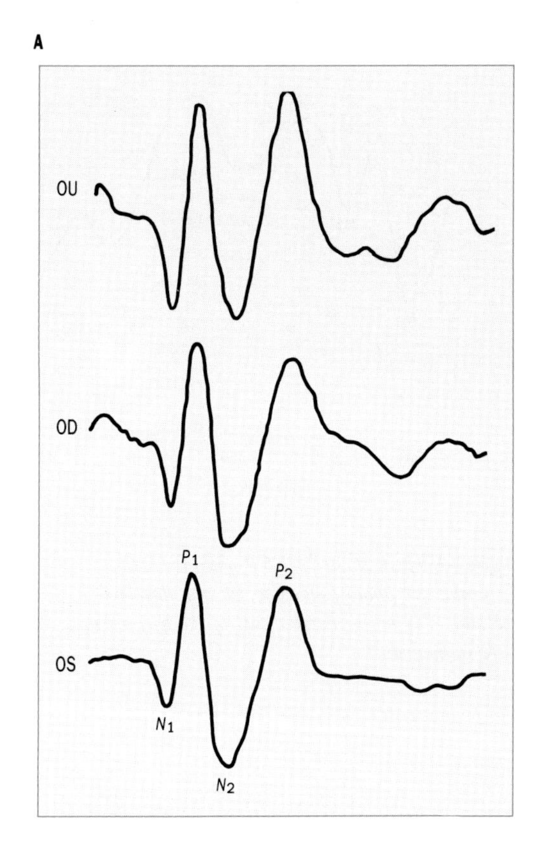

B

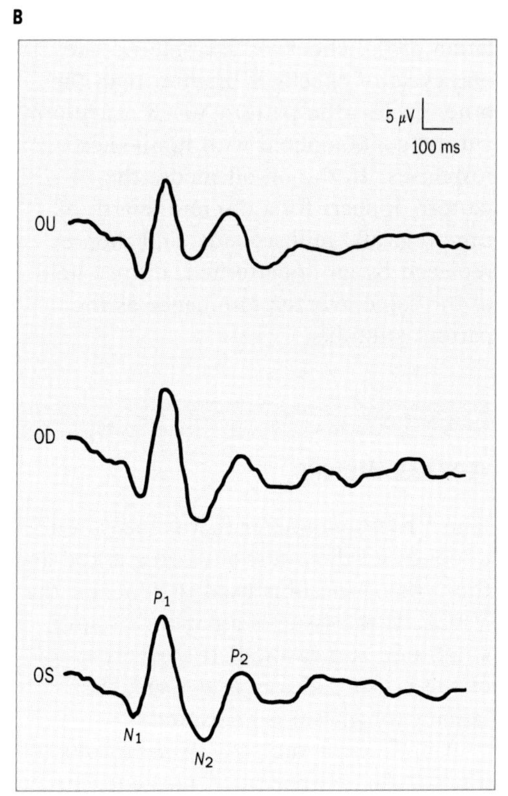

Figure 3-3 *Binocular (OU) and monocular (OD and OS) transient pattern VECP for (A) 12- and (B) 48-minute checks recorded from a visually normal adult. Amplitudes (particularly P_2) are larger for 12-minute checks than for 48-minute checks.*

amplitudes can vary by as much as 25% within and between subjects.[7] Latency is clinically useful because an individual patient's latency can be compared to age-matched norms using statistical probability rules. However, relative amplitude is reliable when comparing interocular differences.[8]

Since latency is a useful clinical parameter, it has been studied extensively in subjects with normal vision. Important factors that affect VECP latency include pupil diameter, age, and refractive error. As shown in Figure 3-4, a decrease in pupil diameter results in a prolongation of P_1 latency, caused by a decrease in retinal illuminance.[9] Latency decreases by nearly 20 milliseconds when pupil diameter increases from 1 to 9 mm. A pupil diameter larger than 4 mm produces a generally constant latency. As shown in Figure 3-5, P_1 latency for large and small checks recorded from subjects between 2 months and 80 years of age[10] decreases exponentially until the 40s and then increases slightly. The functions are the same shape for both check sizes, but absolute latency is longer for small checks. Figure 3-6 shows VECP latency recorded with lenses of varying power from two subjects after cycloplegia: RP, a myope; and JM, a hyperope. The shortest latency was obtained when optical values equivalent to their emmetropic correction were used. In subjects without cycloplegia, minus lenses produce a constant latency until the subject can no longer accommodate.[11] Stimulus factors also affect VECP latency. For example, latency decreases with increasing pattern size,[12-14] increasing luminance, and increasing contrast.[15]

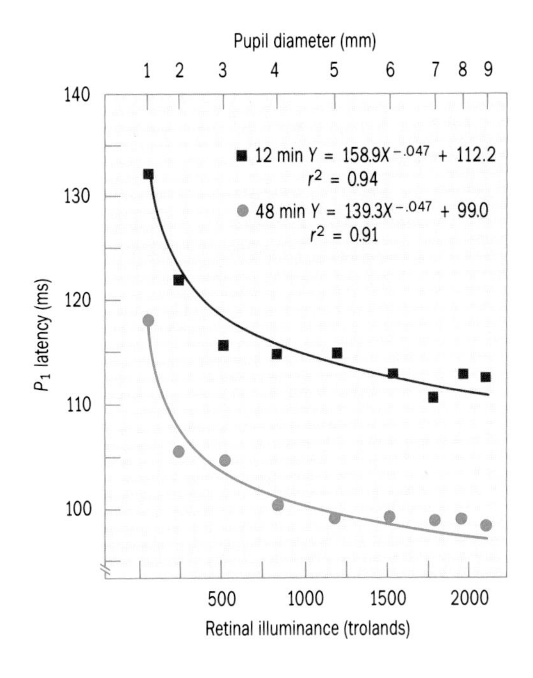

Figure 3-4 *VECP latency in a normal subject for 12- and 48-minute checks as a function of pupil diameter (upper scale). The subject's eye was dilated with 2.5% phenylephrine and tested with artificial pupils ranging in diameter from 1 to 9 mm. The lower scale shows the effective retinal illuminance for each pupil diameter. The data are best fit by a hyperbola; respective equations and coefficients of determination are shown in the figure.*

Reprinted by permission from Sokol S, Domar A, Moskowitz A, et al: Pattern evoked potential latency and contrast sensitivity in glaucoma and hypertension. In: Spekreijse H, Apkarian PA, eds. Electrophysiology and Pathology of the Visual Pathways. *Hague: Dr W Junk BV Publishers.* Doc Ophthalmol Proc Ser *1981;27:79–86.*

Figure 3-5 P_1 *latency as a function of age for (A) large (48- and 60-minute) checks and (B) small (12- and 15-minute) checks. Note that latency for small checks is longer than for large checks across the life span.*

Reprinted by permission from Moskowitz A, Sokol S: Developmental changes in the human visual system as reflected by the latency of the pattern reversal VEP. EEG Clin Neurophysiol *1983;56:1–15.*

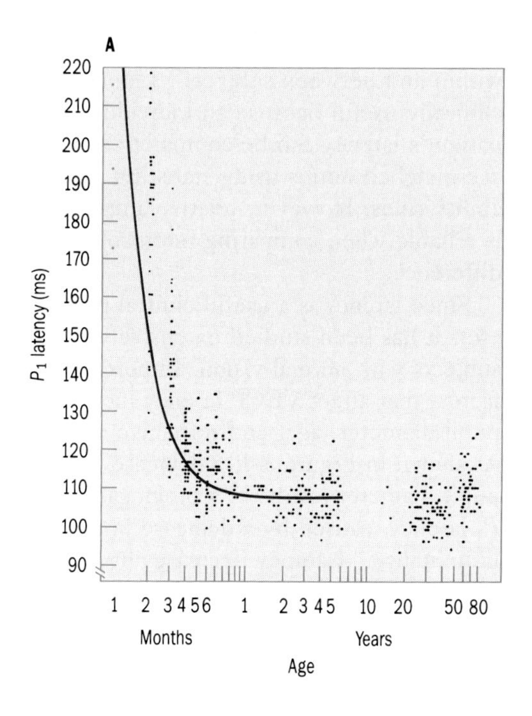

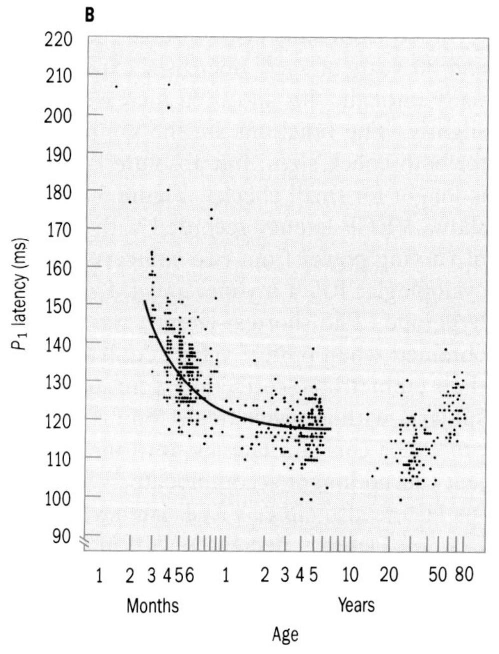

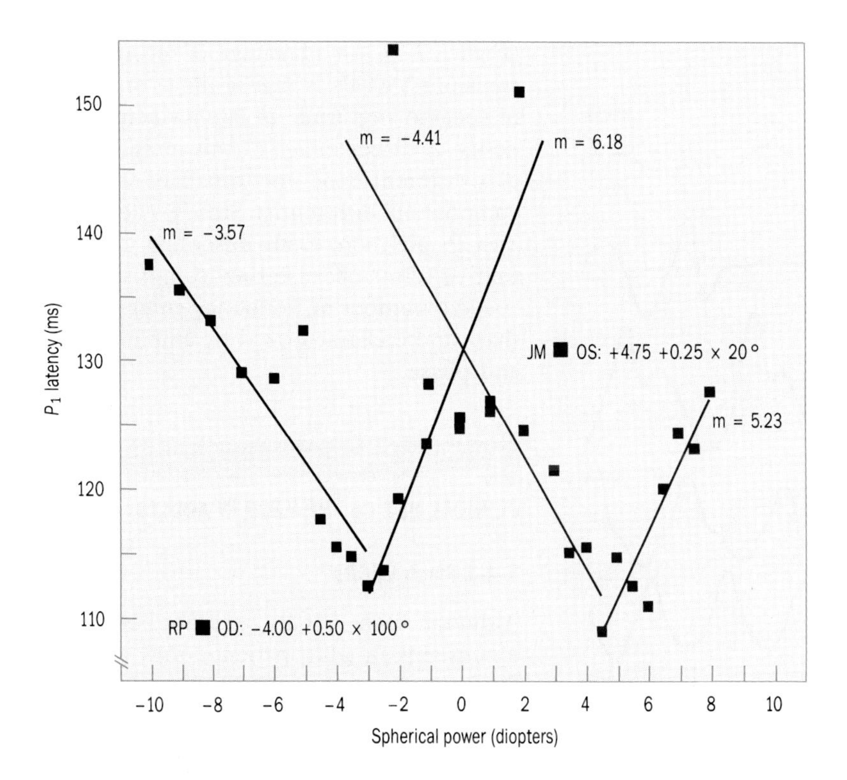

VECPs are either transient or steady-state. A transient VECP is produced when substantial intervals elapse between stimulus presentations, allowing the brain to regain its resting state. Steady-state VECPs are produced when the brain does not regain its resting state between stimulus presentations. A practical distinction between the two is that transient VECPs are evoked by low stimulus rates while steady-state VECPs are evoked by rapidly repetitive stimuli. The temporal rate at which VECPs change from transient to steady-state varies, depending on a variety of factors, but in general the transition occurs at 6–10 presentations per second

Figure 3-6 *P_1 latency as a function of spherical power for 12-minute checks for a myope (subject RP, black squares) and a hyperope (subject JM, red squares). Both subjects underwent cycloplegia. Note that the shortest latency is found for each subject's best optical correction (m-slope).* *Reprinted by permission from Sokol S, Moskowitz A: Effect of retinal blur on the peak latency of the pattern evoked potential.* Vision Res *1981;21:1279–1286. Copyright 1981, Pergamon Press PLC.*

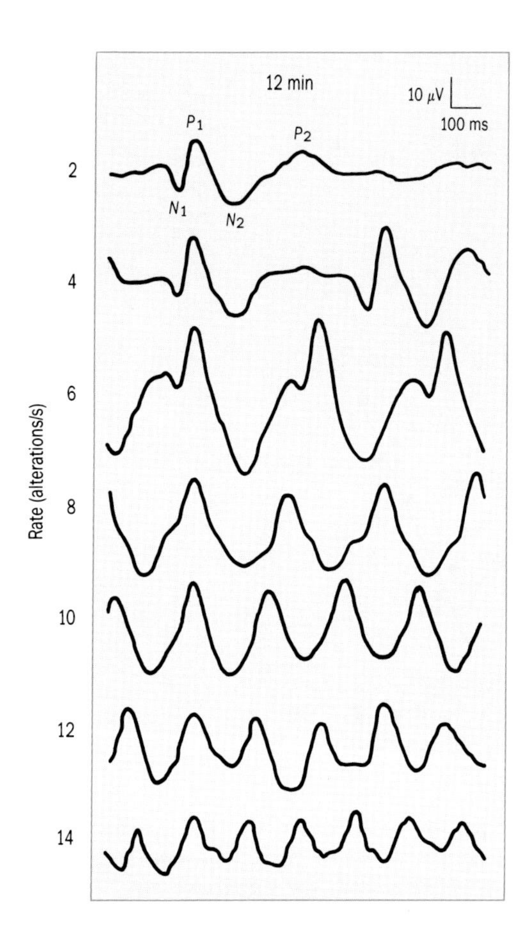

Figure 3-7 *Pattern-reversal VECPs in response to 12-minute checks at different alternation rates. The rate, in alternations per second, is shown to the left of each waveform. The change from transient VECPs to steady-state VECPs is gradual, occurring somewhere between 6 and 10 reversals per second.*

Reprinted by permission from Sokol S: Visual evoked potentials. In: Aminoff MJ, ed. Electrodiagnosis in Clinical Neurology. *2nd ed. New York: Churchill Livingstone; 1986:441–466.*

(Figure 3-7). An advantage of recording transient VECPs is that it allows analysis of specific amplitude or latency components. A disadvantage of this method is that different experimenters may measure components differently. Steady-state VECPs are more easily described in quantitative terms because they comprise a specific number of frequency components that can be characterized by amplitude and phase.

3-4

CLINICAL USE OF THE VECP IN ADULTS

3-4-1 Flash VECPs

Although flash VECPs have not been used extensively in adult patients, they are useful for testing patients with dense media opacities.[16] Odom et al,[17] using a technique described earlier by Weinstein,[18] recorded flash VECPs from cataract patients who had varying degrees of visual acuity reduction. Using 10-Hz flashes, they found a good correlation between the VECP prediction of postoperative acuity and the postoperative subjective visual acuity. The flash VECP has also been useful in the selection and timing of vitreous surgery in eyes with diabetic vitreous hemorrhage.[19,20] Scherfig et al[21] recorded flash VECPs from diabetic patients prior to vitrectomy and found that patients with latencies of less than 100 milliseconds were significantly more likely to have improved vision after vitreous surgery than were patients with latencies longer than 100 milliseconds.

3-4-2 Pattern VECPs

The VECP reflects activity of the sensory visual pathways, beginning at the retina and ending at the occipital cortex. Thus, pathology anywhere along the visual pathways produces abnormal VECPs. A summary of VECP amplitude, latency, and waveform changes in disorders of the optic nerve and visual pathways is given in Table 3-1. Some of the more important findings are described below.

3-4-2-1 Retinal Disease VECP latency is prolonged in patients with central serous retinopathy.[22-25] However, with recovery of visual function, latency returns to normal.[23] This finding is in contrast to latency abnormalities in optic neuritis (see below), which persist even after acuity has returned to normal. Johnson et al[26] tested patients with optic neuritis, maculopathy, macular holes, macular cysts, and lamellar holes. They found that patients with lamellar holes had normal latency, but patients with macular cysts and macular holes had prolonged P_1 latencies. They also found a greater prolongation of P_1 latency in patients with optic neuritis than in patients with macular holes. Johnson et al explained these prolonged VECP latencies on the basis of amplitude summations that occur with half-field VECPs (see Figure 3-1) rather than to a true conduction delay such as that Fukui et al found in eyes with optic neuritis.[27]

3-4-2-2 Optic Neuritis P_1 latency of the pattern-reversal VECP is delayed in patients with optic neuritis.[28] This prolongation persists even after visual acuity returns to normal. While there have been a large number of confirmatory reports of prolonged latency in optic neuritis since the original finding by Halliday et al,[28] abnormal VECP latency is not pathognomonic of optic neuritis. Nevertheless, the pattern VECP can be useful in confirming a previous attack of optic neuritis or retrobulbar neuritis.

3-4-2-3 Multiple Sclerosis The diagnosis of multiple sclerosis (MS) depends on clinical or laboratory evidence for the presence of multiple lesions in the central nervous system. The optic nerve is one of the most common sites to be involved. Measurement of the latency of the response evoked by pattern stimulation provides an objective means of identifying lesions of the visual pathways, even when they are subclinical.[29,30] Accordingly, when patients presenting with clinical evidence of a single lesion of the nervous system (especially a lesion below the level of the foramen magnum) are being evaluated, VECP studies provide a means of establishing the presence of multiple lesions by permitting the detection of clinically silent pathology in the visual system. Abnormal responses to pattern stimulation in patients without a history of optic nerve involvement and without eye symptoms at the time of testing indicate that the VECP can reveal subclinical involvement of the visual pathways in patients with MS.[31]

The high diagnostic yield of VECPs in patients with definite MS has been confirmed repeatedly, with a range usually varying between 70% and 97%. The percentage of abnormal VECPs is lower

TABLE 3-1

VECP Abnormalities

Disorder	Amplitude	Latency	Morphology
Optic Nerve and Visual Pathway			
Optic neuritis	A/N	+++	N
Ischemic optic neuropathy	−−	+	A
Toxic amblyopia	−	N	A
Dominant optic atrophy	−	N/+	?
Leber's optic atrophy	−−	+	A
Optic nerve hypoplasia	−−	+	A
Glaucoma	N/−	N/−	N
Optic disc drusen	−	+	A
Papilledema	N/−	N/+	N/A
Tumors (anterior visual pathways)	−	+	A
Neurologic Disorders			
Multiple sclerosis	N/−	+++	N/A
Vitamin B_{12} deficiency	N	+	N
Congenital nystagmus	−	+	A
Parkinson's disease	N/−	N/++	N
Migraine	N	+	N/A
Down syndrome	−	N	?
Cortical blindness	N/−	N	N
Occipital lobe lesions	−	+	A
Huntington's disease	−	N	A
Friedreich's ataxia	−	++	A
Hereditary spastic ataxia	N	N/+	N
Nonspecific recessive ataxia	N	N	N
Charcot-Marie-Tooth disease	?	+	N
Phenylketonuria	N	+	N

N = Normal; A = Abnormal; ? = Information not available; +(−) = Mild increase (or decrease); ++(−−) = moderate increase (or decrease); +++(−−−) = severe increase (or decrease).

Reprinted by permission from Sokol S: Visual evoked potenitals. In: Aminoff MJ, ed. Electrodiagnosis in Clinical Neurology. *2nd ed. New York: Churchill Livingstone; 1986:441–466.*

among patients in the category of probable MS and lowest among those with possible MS, commonly ranging between 20% and 40%. The incidence with which VECPs of prolonged latency are encountered in patients who have had a single acute neurologic episode of uncertain etiology, but which may be the first clinical manifestation of MS, is less clear. In one study,[32] VECP abnormalities were found in 25% of patients with an acute spinal cord lesion and in 46% of those with an isolated brain-stem lesion; but in another VECP study,[31] latency was not prolonged in any patient with an isolated brain-stem disturbance.

Attempts to increase the diagnostic yield of VECP recording among individuals suspected of having MS have employed a variety of different means—such as altering the check size, field size, luminance, orientation, or alternation rate of the stimulus or changing body temperature—but such approaches have not met with general acceptance.

The pathophysiologic basis of the VECP changes described above remains uncertain. Direct experimentation has shown that a complete conduction block may result from extensive demyelination of central nerve fibers, while conduction velocity is slowed in less severe and less extensive demyelinating lesions. This finding suggests that delay in the latency of VECPs may reflect a reduced conduction velocity in damaged visual nerve fibers, although some of the prolongation in latency may also be due to delay at the cortical or retinal level. Amplitude changes presumably reflect, in large part, a complete conduction block in damaged fibers.

3-4-2-4 Compressive Lesions The latency of the VECP is prolonged by compressive lesions, but only at an early stage. Moreover, even when there is an increase in latency, it is generally much less in patients with compressive lesions than in those with demyelinating disease. For example, Halliday et al[33] found that latency was not delayed more than 20 milliseconds beyond the upper limit of the normal range in patients with compressive lesions, while in optic neuritis associated with MS the mean delays were between 35 and 45 milliseconds and actual delays in individual cases ranged up to 100 milliseconds. In addition, VECPs in patients with compressive lesions showed a much higher incidence of waveform abnormalities (abnormally shaped $N_1P_1N_2$ complex) than in patients with demyelinating disorders. Asymmetry of the VECP was especially characteristic of patients with a tumor arising in the region of the sella turcica.[34] Such changes cannot, however, be regarded as pathognomonic of the underlying lesion. In several cases, abnormal responses were recorded from the clinically normal eye, suggesting that the VECP may permit detection of subclinical damage to optic nerve fibers in this clinical context as well as in patients with demyelinating disease.

3-4-2-5 Functional Disorders: Malingering and Hysteria Since the pattern VECP reflects the ability to see, it is useful, with some caveats, for testing patients who claim they cannot see but have no obvious pathology.[35] For example, Figure 3-8

A

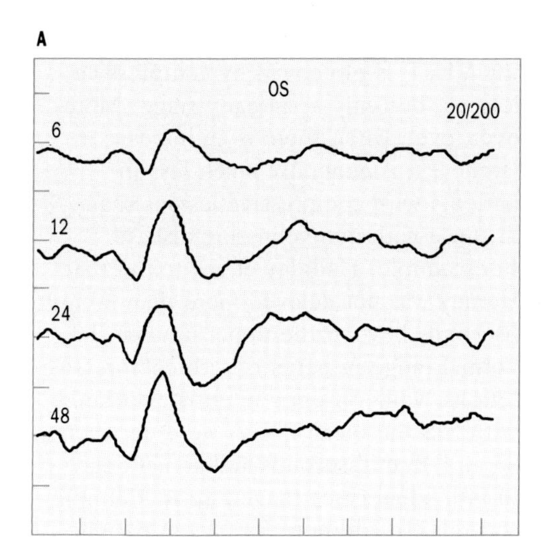

B

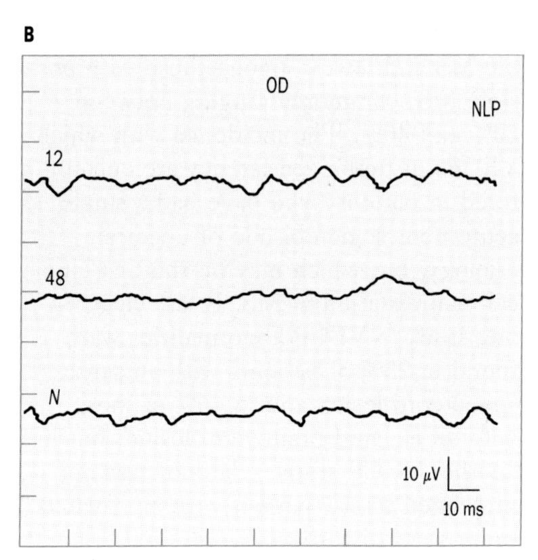

Figure 3-8 *Pattern VECPs recorded from a 55-year-old patient who had aphakia, macular degeneration, and peripheral retinal degeneration in the right eye. (A) Examination of the left eye was normal, yet the patient claimed a visual acuity of 20/200. VECPs recorded for checks of 6–48 minutes show large amplitude, normal latency signals. (B) Signals from the right eye were nondetectable and not significantly different from noise (N). NLP = no light perception.*

shows pattern VECPs recorded from a patient with documented pathology and blindness in the right eye but no pathology in the left eye, yet the patient claimed 20/200 acuity. Clearly, pattern VECPs from the left eye were normal. In this case, a positive finding of a recordable pattern VECP to small checks is helpful in establishing the presence of intact visual pathways. However, absent VECPs in a patient claiming blindness can pose a dilemma, since this result could relate either to pathology or to voluntary suppression of the VECP.

Baumgartner and Epstein[36] reported that 5 of 15 normal subjects could extinguish their pattern VECPs voluntarily, using such strategies as transcendental meditation, concentration beyond the plane of the checks, and ocular convergence. Subsequent studies have shown that deliberate alteration of the VECP can be minimized by using large fields, large

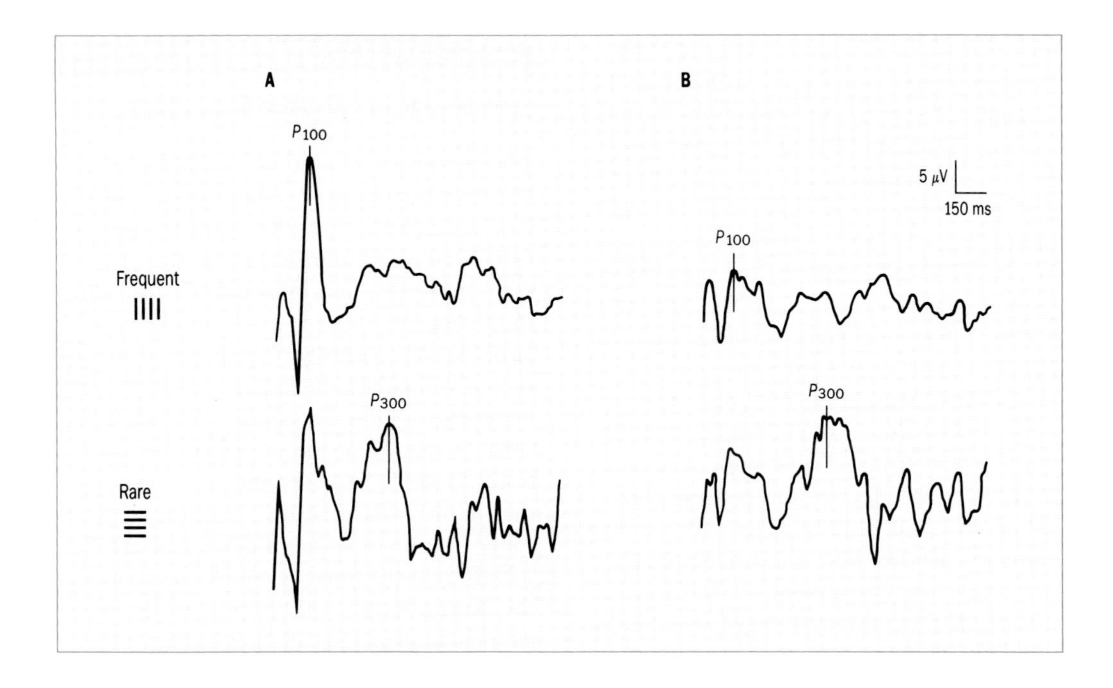

checks, and binocular stimulation.[37-40] Unfortunately, these conditions are not optimal for acuity testing, since the highest correlation of acuity and pattern VECPs is found when small checks and small fields are used. Because the VECP can sometimes be suppressed voluntarily, the pattern VECP should be used with caution in patients suspected of malingering.

The use of long-latency potentials, such as the P_{300} response, that are influenced by sensory stimuli but not evoked by them can eliminate problems of voluntary suppression.[41,42] The P_{300} response is elicited by unpredictable or infrequent stimuli and cannot be voluntarily suppressed if the subject is attending to the task. Figure 3-9 shows the responses

Figure 3-9 *Pattern VECPs (6-minute gratings) and P_{300} evoked potentials obtained under two conditions: attend and blur. (A) In the attend condition, the subject could easily distinguish between the vertical and the horizontal gratings and correctly counted the number of times a horizontal grating appeared. (B) In the blur condition, +20-D lenses were placed in front of the subject's eyes. The subject reported that, even though the gratings were difficult to see, she could still distinguish the different orientations. In this condition, the P_{100} component of the pattern VECP was severely attenuated, but the P_{300} component in response to the rarely presented orientation was still present.*

Reprinted by permission from Sokol S: Visual evoked potentials. In: Aminoff MJ, ed. Electrodiagnosis in Clinical Neurology. *2nd ed. New York: Churchill Livingstone; 1986:441–466.*

obtained when the subject viewed a TV monitor on which vertical and horizontal gratings were presented randomly. Vertical gratings appeared 80% of the time (frequent), and horizontal gratings appeared 20% of the time (rare). The subject was instructed to count the horizontal gratings. When the stimuli were in focus (attend), large P_1 waves were present for both horizontal and vertical gratings and a large P_{300} wave occurred for the rarely presented horizontal gratings. The subject then viewed the gratings through a 20-D positive lens (blur). While no longer able to resolve the edges of the gratings, the subject was still able to detect orientations correctly. Under these conditions, P_{100} was attenuated (as predicted), but P_{300} remained. Such findings suggest that subjects who voluntarily blur or attempt to tune out a pattern stimulus may be unable to suppress their P_{300} response, although they may attenuate their P_{100} amplitude.

Another solution to the problem of testing patients suspected of voluntarily reducing their VECPs is to present pattern onset-offset stimuli randomly. If the subject is not aware of when the stimulus is going to appear, it is more difficult to be deceptive.

3-5

CLINICAL USE OF THE VECP IN CHILDREN

The infant visual system continues to develop after birth.[43] While the retinal rods are nearly adult-like at birth, the foveal cones are immature and the ganglion cells have not moved aside to form the foveal pit.[44,45] Myelination of the optic nerve continues,[46] cell size of the lateral geniculate nucleus increases,[47] and the synaptic junctions and cell types in cortical area 17 of the occipital lobe continue to grow, expand, and develop.[48] As the infant's visual system matures anatomically, there are changes in the pattern VECP as well. Smaller check sizes elicit the largest VECP amplitude, and peak latency shortens for all check sizes, more rapidly for large checks than for small checks[49-51] (Figure 3-10). Infant acuity, measured with the VECP, improves to a Snellen equivalent of 20/20 by 6 months to 1 year of age (Figure 3-11).

3-5-1 Flash VECPs

Flash VECPs are used primarily as a noninvasive means of evaluating the integrity of the sensory pathways of very young infants. The emphasis has been on the use of the flash VECP to evaluate the effect of intraventricular hemorrhages (IVH) on cortical maturation and mental development and to evaluate the integrity of shunts, which are used to control increased intracranial pressure in posthemorrhagic hydrocephalus.

Placzek et al[52] recorded flash VECPs from premature infants who ranged from neurologically normal without IVH to neurologically abnormal with grade III IVH. While the time course of the VECP waveform maturation was the same for all infants tested, Placzek et al found a delayed onset of the first major positive wave in the severely ill children. Similar findings have been reported by Kurtzberg.[53] Mushin et al[54] also found abnormal flash VECPs in infants with IVH, but argued that these abnormalities reflected subcortical function, raising the question of whether the flash VECP is useful as an accurate predictor of later visual development. Support for their hypothesis is based on their finding that infants with periventricular leukomalacia (PVL), an infarction of white matter that can affect visual radiations, show normal fixation and tracking in the neonatal period but exhibit abnormal visual behavior after 8 weeks of age, when vision is predominantly cortical. Related to this finding are reports of normal flash VECPs in cortically blind children (see below).

Flash VECP studies of infants with posthemorrhagic hydrocephalus have shown that the latency of the primary positive wave increases linearly as intracranial pressure increases.[55-58] This relationship is affected by a number of factors, however. Guthkelch et al[59] found that latency (1) increases in inverse proportion to the age at which symptoms appear, (2) is significantly longer when the head is enlarged, and (3) is longer still when the patient has ventriculitis. They also found that latency did not change in parallel with changes in ventricular size but did correlate with evidence of brain

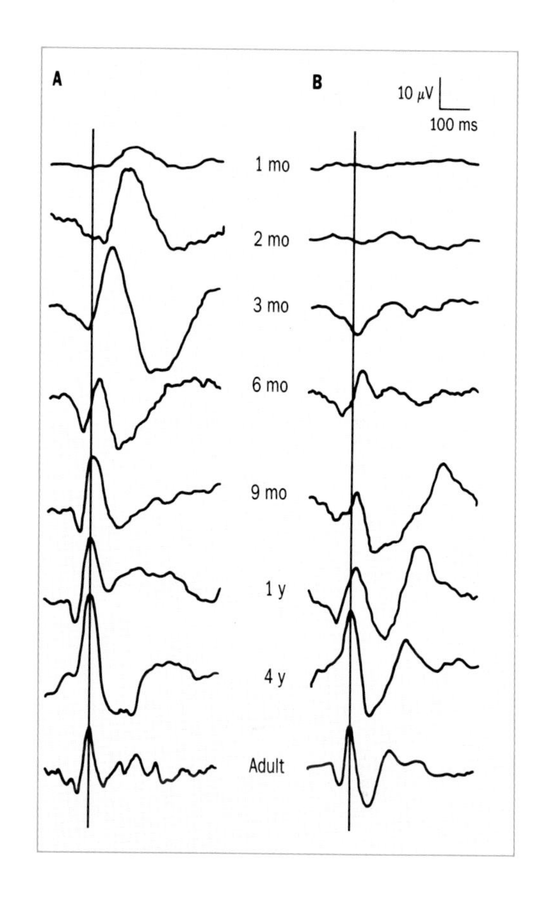

Figure 3-10 *Development of VECP waveform for (A) large (60-minute) and (B) small (15-minute) checks.*

Reprinted by permission from Moskowitz A, Sokol S: Developmental changes in the human visual system as reflected by the latency of the pattern reversal VEP. EEG Clin Neurophysiol *1983;56:1–15.*

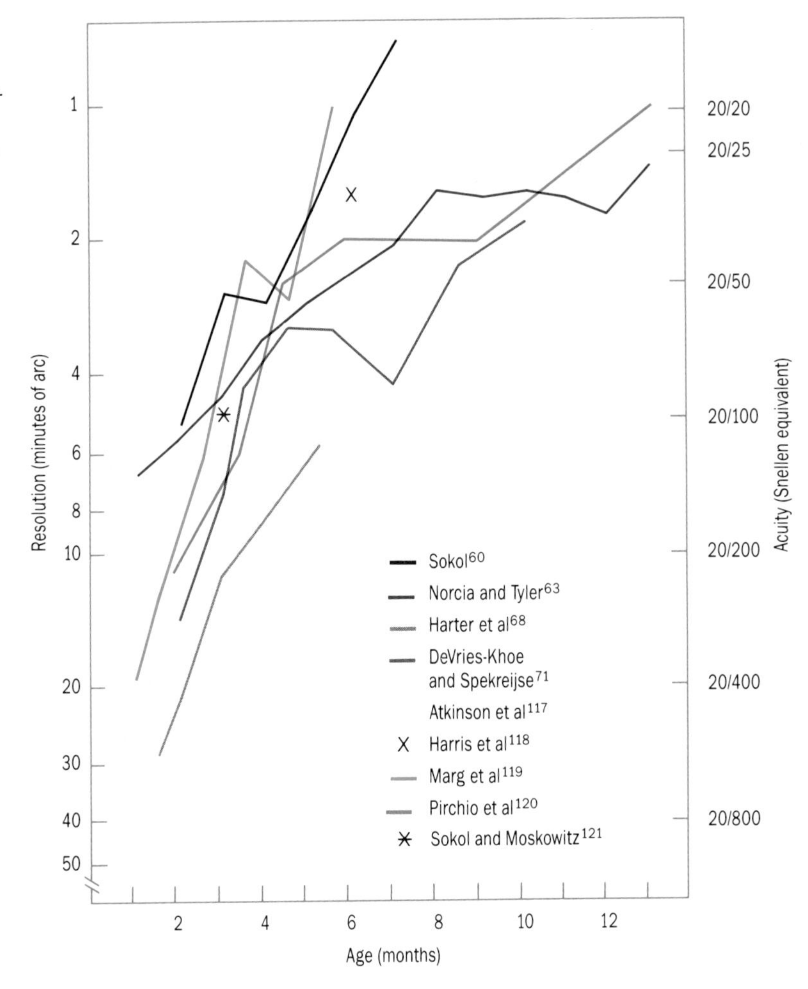

Figure 3-11 *Development of infant visual acuity as measured by the pattern VECP. (Sources for this figure are indicated on the figure.)*

damage. Retarded hydrocephalic infants had abnormally prolonged latencies, while the latencies recorded from developmentally normal hydrocephalic infants were normal. The authors concluded that the value of the flash VECP lies in assessing mental development not in monitoring shunt function.

To evaluate the central nervous system of high-risk infants, Kurtzberg[53] recorded flash VECPs from 79 high-risk infants when they reached 40 weeks conceptional age (CA), which corresponds to term gestation. This allowed comparison of the data from high-risk infants with the data from a normal full-term population. In the group of 79 high-risk infants, 51% had normal flash VECPs at 40 weeks CA. The remaining 49% had abnormalities that consisted of hemispheric wave-shape asymmetries, amplitude asymmetries, and immature or atypical responses. The validity of the flash VECP measures was then assessed by retesting the infants at monthly intervals. The results showed that 92% of the infants with normal flash VECPs at 40 weeks CA had normal VECPs during the first 6 months of life and at 1 year. Abnormalities persisted in more than 50% of the group with wave-shape and amplitude asymmetries and in the group with immature responses, as well as in more than 75% of the group with atypical responses. Standard neurologic and developmental examinations were most likely to be normal in infants with normal term (40 weeks CA) flash VECPs. Flash VECPs can also be useful for testing infants with media opacities (*see Section 3-4-1*).

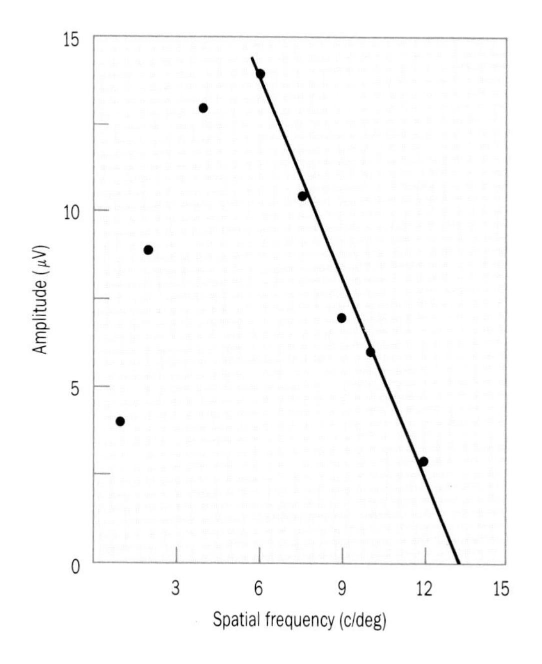

Figure 3-12 *Amplitude (μV) of the transient VECP as a function of pattern size in cycles per degree (c/deg) for a 6-month-old infant. A straight-line fit to the data points is extrapolated to 0 μV. The Snellen equivalent of 13 c/deg, the 0 μV intercept of a straight-line fit from the peak of the function, is 20/45.*

3-5-2 Pattern VECPs

3-5-2-1 VECP Acuity Two basic methods are used to estimate VECP acuity: transient VECP extrapolation[60] and sweep VECP extrapolation.[61-64] In the transient VECP extrapolation method, VECP amplitude is measured for a series of check (or stripe) sizes. This technique is based on the findings of Campbell and Maffei,[65] who demonstrated that there is a strong correlation between psychophysical and VECP estimates of contrast thresholds. To estimate VECP acuity, a straight line is fit from the peak of an amplitude/check-size function and extrapolated to 0 μV. The 0 μV intercept is defined as the subject's acuity (Figure 3-12). This method is useful for

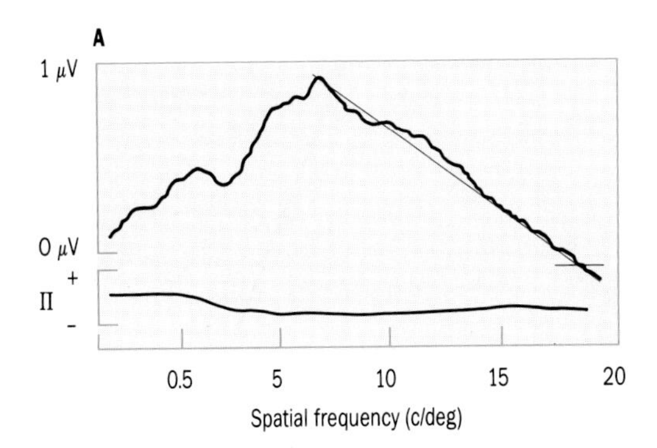

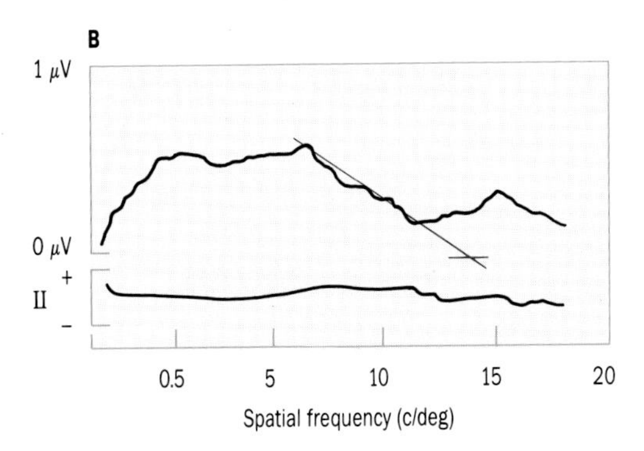

Figure 3-13 *Steady-state sweep VECP grating acuity estimated from the normal (A) and amblyopic (B) eye of an anisometropic amblyope using a sweep VECP technique. The stimulus was a square wave grating alternating at 8 Hz (16 reversals/s). The spatial frequency of the grating was swept linearly from 0.5 to 20 c/deg in 13 s. Amplitude (A) and phase (B) are shown for each eye. Upward movement of phase indicates a lag (delay in latency). A straight line, fit by eye from the peak of each amplitude function, is extrapolated through the 0 μV axis. Grating acuity of the normal eye is 19 c/deg (Snellen equivalent: 20/30); acuity of the amblyopic eye was 14 c/deg (Snellen equivalent: 20/40). The patient's Snellen acuity was 20/20 in the normal eye and 20/50 in the amblyopic eye.*

estimating an infant's binocular acuity, but is unreliable and impractical in a clinical setting, particularly for monocular acuity, because too much time is required to collect the data. A more efficient method for estimating acuity is the sweep technique.

The sweep VECP technique uses a pattern-reversal stimulus of black and white bars that are changed rapidly (less than 30 seconds) in size from wide to narrow widths. At the same time, a narrow-frequency band of EEG information is analyzed with Fourier techniques.[62-64,66] Acuity is estimated by extrapolating the peak of the VECP response function to 0 μV or noise (Figure 3-13). The sweep technique is more expedient than the transient VECP extrapolation method but, in exchange for speed, information from cortical frequencies other than the frequency being analyzed is sacrificed. In spite of these drawbacks, the sweep VECP promises to be an important technique for estimating acuity in pediatric patients.[67]

3-5-2-2 Suprathreshold Vision In addition to VECP acuity, which measures the limit of visual resolution (threshold), the VECP can also measure visual function in response to large (suprathreshold) pattern stimuli. The advantages of this strategy are that the signal-to-noise ratio is large and that measurement of VECP components is easier. Suprathreshold VECPs can be used to monitor the visual development of pediatric patients by comparing their amplitude and latency to age-matched norms.[49,51] Measurement of latency is particularly useful since it is stable within and between age-matched subjects.[7]

Another suprathreshold technique, the measurement of interocular amplitude difference, is useful when studying a patient with monocular visual loss. It is based on the assumption that if a reliable response is obtained for checks of 10–20 minutes (known to test foveal function[3]) and the amplitude is equal between eyes, the subject has normal and equal acuity.[8] This assumption is based on the fact that the peak and the 0 μV intercept of the amplitude check-size function shift, in parallel with each other, to higher spatial frequencies (smaller checks) from birth to 3 years of age.[68-71] After 3 years of age, the 0 μV intercept continues to shift while the peak remains constant at 10–20 minutes of arc.[71]

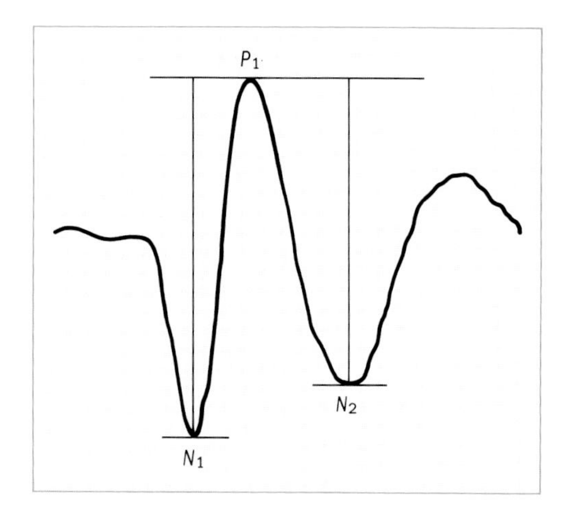

Figure 3-14 *Transient pattern VECP showing N_1–P_1 and P_1–N_2 amplitude components. See text for how the interocular amplitude difference is calculated.*

To calculate the interocular amplitude difference, the amplitude, in microvolts, of N_1–P_1 and P_1–N_2 is measured (Figure 3-14). For both components, the value obtained from the left eye is subtracted from the value for the right eye and the difference is divided by the larger of the two; this calculation produces a normalized "interocular amplitude difference ratio." Acuity between eyes is considered abnormal when *both* N_1–P_1 and P_1–N_2 difference ratios are greater than the normal mean interocular ratio plus 1 standard deviation.[8] It should be emphasized that interocular amplitude measurements are not direct estimates of acuity, but the empirical fact is that abnormal interocular VECP amplitude correlates well with unequal visual acuity (*see the next section*).

3-5-2-3 Amblyopia A traditional clinical technique, such as fixation behavior, can be used to judge which eye is amblyopic in a preverbal child, but it cannot estimate

A

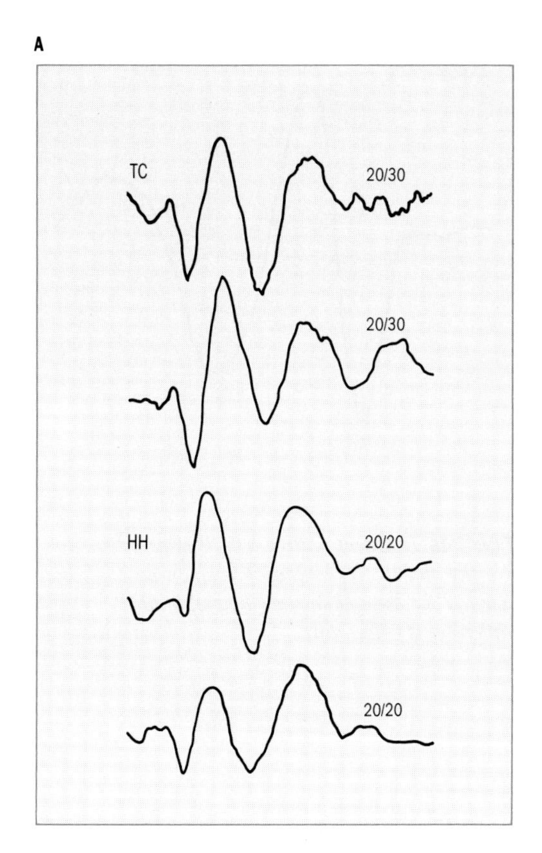

B

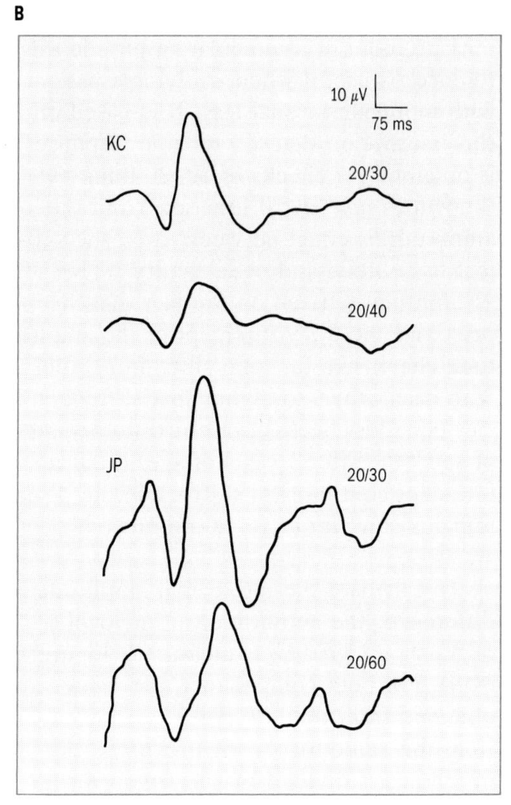

Figure 3-15 *Transient monocular pattern VECPs from (A) two normal patients, TC and HH, and (B) two amblyopic patients, KC and JP. Interocular amplitude differences were abnormal (as indicated by the asterisk) for both N_1P_1 and P_1N_2 for KC and JP, compatible with amblyopia. JP's latency was also abnormal in the amblyopic eye. Snellen acuity measurements confirmed the VECP results.*

Reprinted by permission from Sokol S: Alternatives to Snellen acuity testing in pediatric patients. Am Orthoptic J *1986;36:5–10.*

accurately how well (or poorly) a patient "sees."[72,73] The examiner can only grossly estimate the patient's acuity with this information. The pattern VECP has been shown to be a sensitive detector of amblyopia, particularly when small (<20-minute) checks are used.[62,74-77] Both the sweep VECP technique (Figure 3-13) and the interocular amplitude method detect amblyopia (Figure 3-15). As shown in Figure 3-16, the sensitivity of the interocular method is 100% when the acuity difference between eyes is 3 lines or greater;

sensitivity falls to 50% detection for Snellen acuity differences of 2 lines or fewer. Pattern VECP latency is also abnormal in amblyopia.[76] However, because the effect is small, latency is not as sensitive a marker of amblyopia as amplitude.

Since VECP acuity parallels the improvement in subjective acuity in verbal amblyopes treated with occlusion therapy, it is useful in monitoring acuity in the patched preverbal patient and in detecting occlusion amblyopia caused by patching too long.[78-82] Figure 3-17 compares changes in the interocular amplitude ratio of the pattern VECP, clinical acuity, and fixation preference of an older, verbal patient who underwent occlusion therapy for amblyopia. There is good agreement among the different measures. Figure 3-18 shows changes in the VECP waveforms of an 18-month-old patient with strabismic amblyopia.

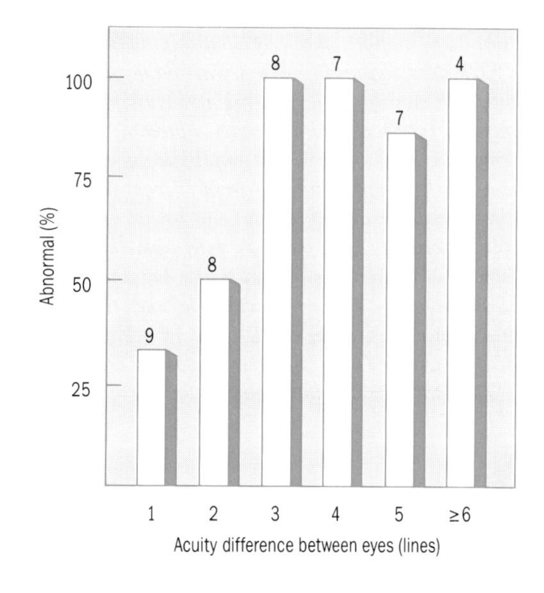

Figure 3-16 *Percent of amblyopes with abnormal VECP interocular amplitude differences as a function of their interocular line difference for Snellen acuity. All but one amblyope with an interocular acuity difference greater than 3 lines had an abnormal interocular acuity VECP difference.*

Published courtesy of Ophthalmology *from Sokol S, Hansen V, Moskowitz A, et al: Evoked potential and preferential looking estimates of visual acuity in pediatric patients.* Ophthalmology *1983;90:552–562.*

3-5-2-4 Refraction A patient's refractive error can be measured electrophysiologically by placing lenses of different power in front of the patient's eye(s) and recording the pattern VECP.[83,84] The lens producing the largest-amplitude and shortest-latency VECP gives clearest vision to the patient. While retinoscopy is clearly the method of choice for measuring a patient's refractive error, it gives only the optical correction necessary for a clear retinal image—it cannot estimate acuity. Thus, the VECP recorded in conjunction with retinoscopic correction is a useful adjunct. If VECP amplitude increases and/or latency decreases with correction, then the correction has likely improved the

Figure 3-17 *Comparison of (A) clinical and (B) VECP data obtained from a 3-year-old accommodative esotrope while undergoing occlusion therapy. Interocular visual acuity difference (squares), interocular VECP amplitude difference (circles), and fixation preference (triangles) were monitored during the course of treatment. The left ordinate of A shows the interocular difference (in octaves) for clinical acuity (an octave is a doubling or halving of acuity; ie, one octave is a change of 20/20 to 20/40). The left ordinate of B shows the VECP difference: equal (0) or abnormal (+, −). Values greater than 0 indicate better clinical acuity and larger VECP amplitudes in the right eye; values less than 0 indicate better performance by the left eye. The right ordinates of both A and B show binocular fixation pattern gradations; upward numbers indicate a fixation preference for the right eye, and downward numbers indicate a fixation preference for the left eye. A grade of 1 indicates a strong preference for one eye over the other; a grade of 5 indicates that the patient alternates. The abscissa shows time after the initial examination of the patient. The vertical line indicates the time of surgery.*

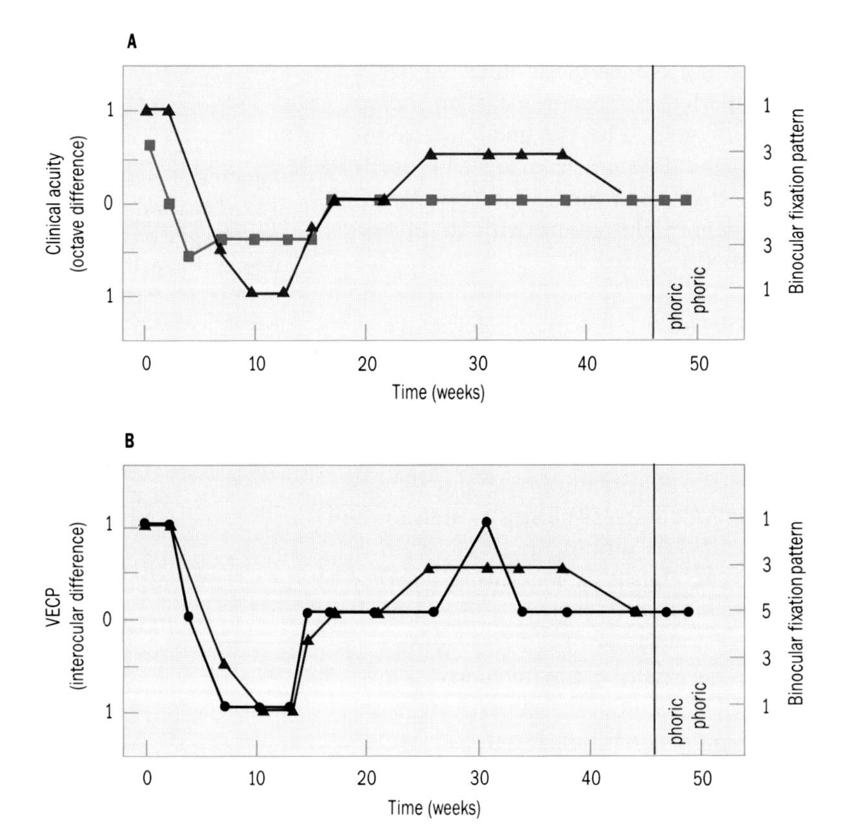

A

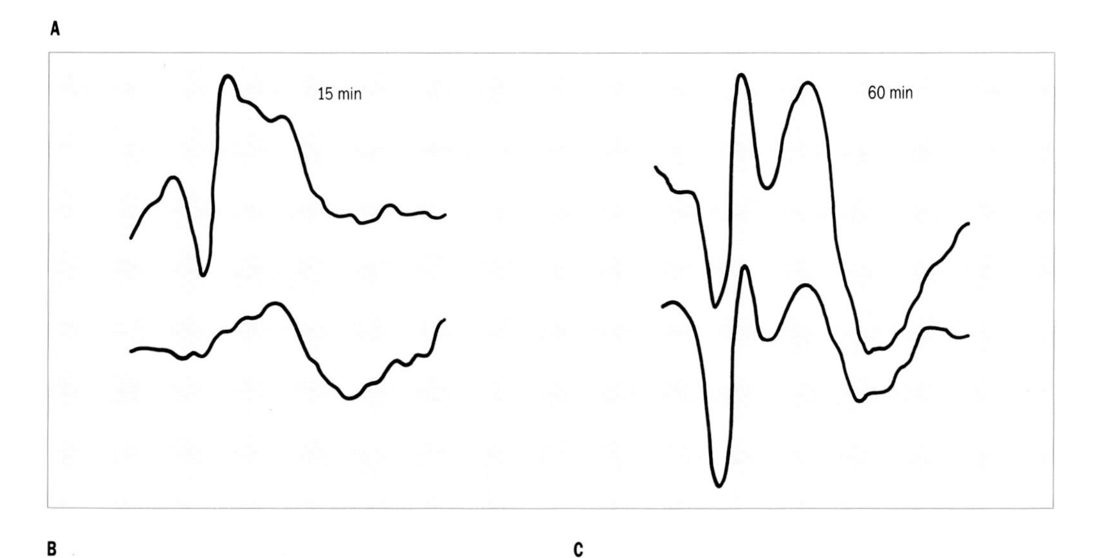

B

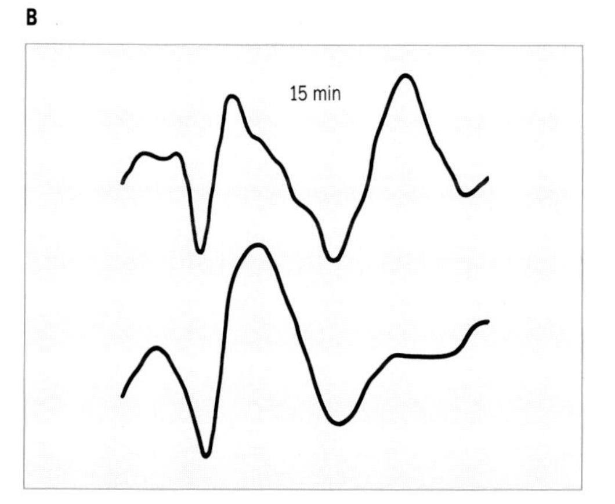

C

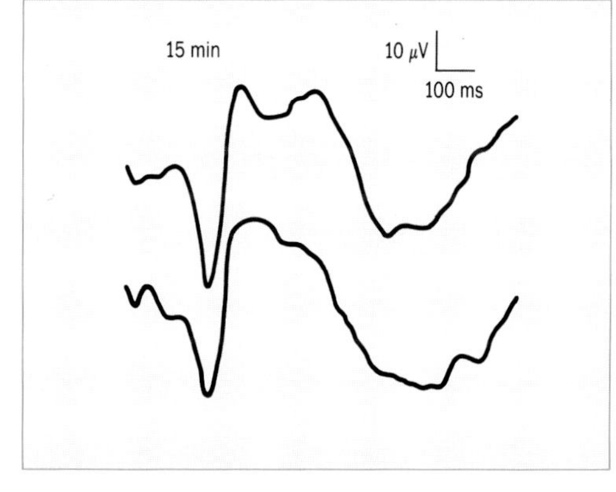

Figure 3-18 *Monitoring of occlusion therapy using the interocular VECP amplitude difference. The records were obtained from an 18-month-old patient with a 25-PD exotropia. (A,* **left***) Records obtained before occlusion therapy using 15-minute checks. (A,* **right***) Records show that amblyopia is undetected when large (60-minute) checks are used. (B) Records obtained after 3 weeks of full-time occlusion. (C) Records obtained when the patient was 3 years old and had equal Allen acuity.*

Reprinted by permission from Sokol S: Alternatives to Snellen acuity testing in pediatric patients. Am Orthoptic J *1986; 36:5–10.*

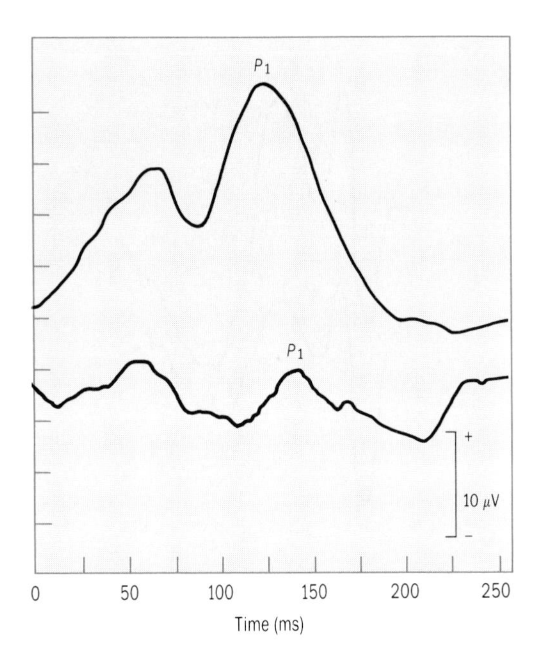

Figure 3-19 *Binocular pattern VECPs (15-minute checks) recorded without (lower) and with (upper) retinoscopically determined correction in a 4-year-old myopic patient. Note the reduced amplitude and prolonged latency without correction.*

patient's acuity. Figure 3-19 shows binocular VECPs recorded from a myopic patient with (above) and without (below) retinoscopically determined correction. The VECP amplitude is larger with correction, indicating better visual resolution at both the retinal and the cortical level. This example highlights the importance of being certain that patients referred for VECP testing have had a refraction and are wearing any necessary correction.

3-5-2-5 Oculomotor Disorders Patients with intact sensory pathways but eye movement abnormalities may appear visually impaired because they cannot track a moving target. One example is congenital ocular motor apraxia, an inability to elicit horizontal saccadic or pursuit eye movements.[85] Infants with poor behavioral visual function, full random eye movements without nystagmus, contraversive deviation of the eyes on rotation of the head, and normal VECPs likely have congenital ocular motor apraxia.[86] By 6 months of age, these patients exhibit compensatory head thrusts, abrupt turnings of the head that allow them to fixate and obtain good vision. Pattern VECPs recorded before the development of these compensatory behaviors can help to confirm the diagnosis. The pitfalls encountered when recording pattern VECPs in patients with nystagmus are discussed in Section 3-5-2-9.

3-5-2-6 Cortical Blindness A patient with absent vision, normal pupillary responses, no nystagmus, and a normal fundus may be cortically blind.[87,88] However, before a definitive diagnosis of cortical blindness is made, the patient should have a complete retinal examination, including flash ERGs. Since there is evidence in "cortically blind" patients of residual visual function,[89] Jan et al[90] suggest a less restrictive term: "cortical visual impairment." A diagnosis of cortical visual impairment should not be made until VECPs are nonrecordable over several test sessions.

There are conflicting reports regarding the implications of normal evoked potentials in patients who are behaviorally blind. Bodis-Wollner et al[91] recorded flash and pattern VECPs from a patient who had been blind for 4 years as a result of an acute febrile illness. The patient's flash VECPs were normal. When gratings were used, clear responses were obtained to low spatial frequencies (0.9–3.7 cycles/degree) but there were no detectable signals to higher spatial frequencies (5.0 and 6.0 cycles/degree). Computed tomographic (CT) scans revealed that the patient's striate cortex was destroyed. Regan et al[92] recorded flash VECPs from a blind 4-month-old infant who later showed complete visual recovery. Prior to any improvement in the patient's vision, flash VECPs were recordable but were less complex and of longer latency than his visually normal twin brother's VECP. As the patient's vision returned, his flash VECP showed increases in the number of components and a shortening of the latency of these components. More recently, Kupersmith and Nelson[93] recorded flash and pattern VECPs from 16 infants with cortical blindness due to a variety of causes. While the investigators were unable to obtain measurable pattern VECPs in any of the infants, they found recordable but abnormal flash VECPs in 12 of 13 infants who recovered visually. Unlike Regan et al,[92] they did not obtain follow-up VECPs from their patients.

Sokol et al[94] tested a patient who appeared to be blind from head trauma. Two CT scans did not show occipital lobe injury; instead, there was probable parietal lobe damage. Figure 3-20A shows normal VECPs obtained on the first visit when the patient's clinical findings and preferential-looking behavior indicated blindness. Recordable signals were obtained to all check sizes, and the peak latency of P_1 was normal for her age. Figure 3-20B shows monocular VECPs obtained 8 weeks later, when the patient fixed and followed objects and was visually attentive to her surroundings. When tested the second time, the binocular amplitudes (not shown) were larger for all check sizes and the increase was within the range of normal retest variability. P_1 latencies were unchanged from the first test. Monocular testing for small and large checks indicated equal amplitude signals for the two eyes; P_1 latencies were normal for each eye. Preferential-looking results indicated visual acuity of 20/60 in each eye.

Kupersmith and Nelson[93] suggest that the patient's age at the time of brain damage may be important; younger patients may have a more plastic visual system and be more apt to recover. It is noteworthy that Regan's patient, the patient tested by Sokol et al, and the patients tested by Kupersmith and Nelson were younger than 2.5 years old when they lost vision.

Figure 3-20 *Pattern VECPs recorded from a child who appeared blind as a result of head trauma and later regained normal visual acuity as determined clinically and with preferential-looking techniques. (A) Normal binocular VECPs obtained when the patient appeared blind. Clinically, the patient did not fix, follow objects, or respond to visual threat; her pupils responded briskly to light, and the patient responded equally in all visual fields to very bright lights by closing her eyes. Preferential-looking testing was unsuccessful; the patient would not fix or track the largest gratings. (B) Monocular VECPs obtained after the patient had recovered visually; preferential-looking acuity was 20/60 in each eye. VECPs were of equal amplitude and normal latency in each eye.*
Reprinted by permission from Sokol S, Hedges, TR Jr, Moskowitz A: Pattern VEPs and preferential looking acuity in infantile traumatic blindness. Clin Vision Sci *1987;2:59–61. Copyright 1987, Pergamon Press PLC.*

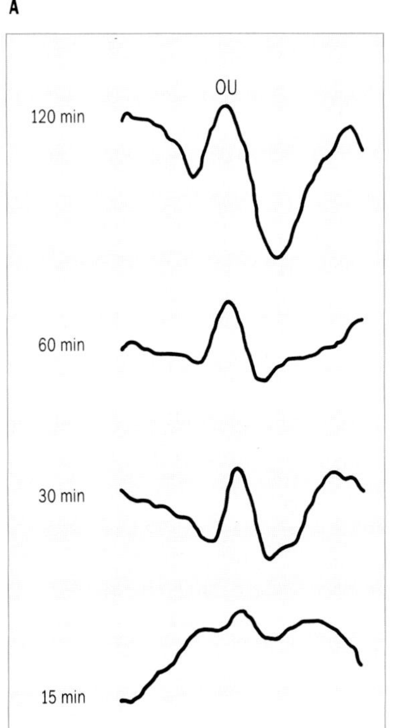

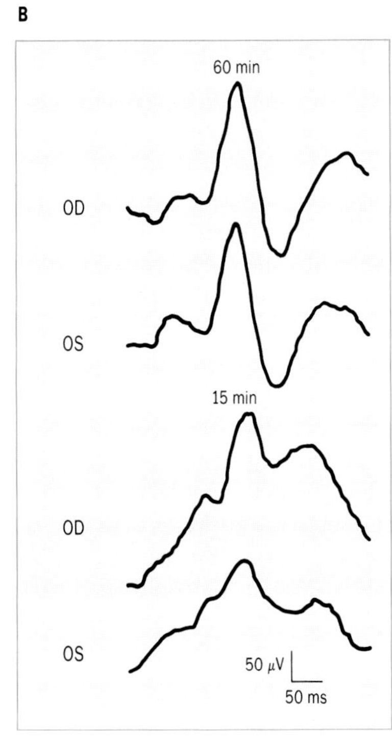

3-5-2-7 Delayed Visual Maturation An otherwise normal infant who is visually inattentive may be suffering from delayed visual maturation.[95-100] These infants fail to fixate and follow, show normal pupillary responses, have no nystagmus, have no structural abnormalities of the eye, and have a normal flash ERG. Eventually, the infants show normal visual behavior, in contrast to infants with cortical visual impairment. Although the VECP is not prognostic in delayed visual maturation, it can be used to monitor visual development. In delayed visual maturation, a systematic improvement in the amplitude and latency of the VECP is seen, as opposed to the repeatedly abnormal or absent VECPs that occur with cortical visual impairment. Possible causes of delayed visual development include delayed retinal development, delayed myelination of the visual pathways, and delayed dendritic and synaptic growth in the visual cortex. Delayed visual development can also be associated with delays of other sensory and motor modalities.[100]

3-5-2-8 Functional Disorders: Malingering and Hysteria A child may claim to have poor acuity without any observable pathology. Frequently, the patient is a female 10–13 years old with complaints about poor vision in school, a condition referred to as "schoolgirl amblyopia."[101,102] A finding of large amplitude, normal latency responses for small (20-minute) checks in these patients is compatible with an intact visual pathway and usually rules out pathology. Abnormal responses may indicate pathology. I have recorded abnormal VECPs from children who were thought to be malingering, but were later found to have a tumor of the anterior visual pathways or optic neuropathy. However, as discussed in Section 3-4-2-5, abnormal or absent responses to small checks may not always indicate pathology. Whether young children are capable of voluntarily suppressing the VECP is open to question; in my experience, they are not.

3-5-2-9 Neurologic Disease Evaluating visual function in neurologically impaired children is particularly difficult. These children have difficulty eliciting motor responses, so behaviorally based tests such as fixation and following with the head and eyes are undependable. Pattern VECPs circumvent the need for discrete motor responses, but the data should be interpreted cautiously. Measuring VECP acuity in a neurologically afflicted patient is different from measuring acuity in a patient with only an eye abnormality. Transient VECP components in patients with central nervous system (CNS) disease are often characterized by changes in polarity and latency, which produce "VECP acuity" data that are difficult to interpret or misleading. Skarf et al[103] reported that they obtained sufficiently reliable VECP responses to estimate an upper limit of visual acuity from 50% to 65% of a group of neurologically impaired children between 7 months and 5 years of age.

Many patients with neurologic disease have nystagmus, which can affect the VECP amplitude in the absence of pri-

A

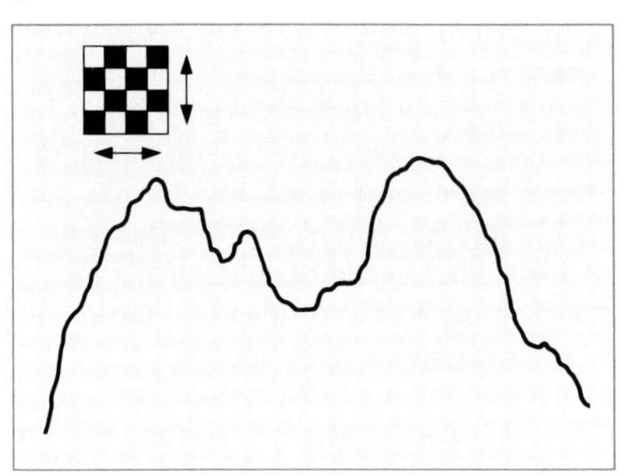

B

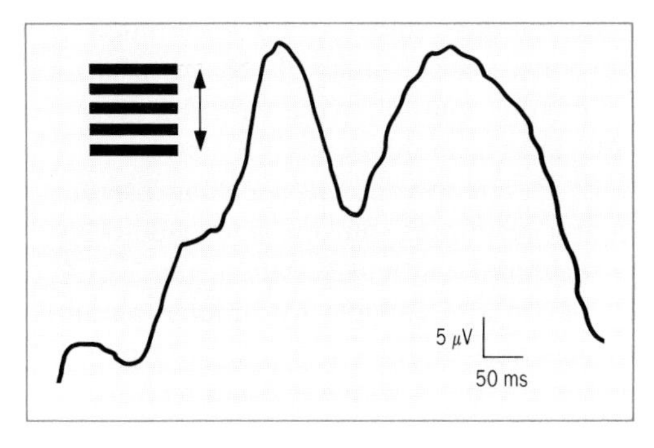

Figure 3-21 *Pattern VECPs recorded from an oculocutaneous albino using (A) phase-reversed checks and (B) horizontal stripes. Preferential-looking visual acuity was 20/170.*

Reprinted by permission from Sokol S: Visual evoked potentials. In: Aminoff MJ, ed. Electrodiagnosis in Clinical Neurology *2nd ed. New York: Churchill Livingstone; 1986:441–466.*

mary sensory impairment. VECP acuity estimates in such patients are often poorer than behavioral estimates, particularly with pattern-reversal stimuli.[8,104] There are several solutions to testing patients with nystagmus. One is to record VECPs with horizontal gratings (Figure 3-21), since they produce less retinal blur on the horizontally moving retina and therefore larger amplitude responses than do checks or vertical gratings. A second solution is to record VECPs with pattern-onset stimuli. Creel et al[105] found large pattern-onset responses in patients with nystagmus and relatively good acuity, while their pattern-reversal responses were nondetectable. Creel et al also found that pattern-onset stimuli were more effective than pattern-reversal stimuli in the detection of chiasmic misrouting in albinism. Finally, instead of VECPs, preferential-looking techniques can be used.[106]

VECPs recorded from hydrocephalic patients who have undergone cerebrospinal fluid (CSF) shunting procedures should also be analyzed cautiously. A prolongation in pattern VECP latency may indicate a malfunctioning shunt rather than poor vision.

3-5-3 Other Techniques

Ideally, a variety of tests should be used to evaluate a preverbal patient's vision. The pattern and flash VECP, ERG, optokinetic response drum, and preferential-looking tests[106,107] are all useful. As more laboratories use these tests and gain clinical experience, the test best suited for a particular visual problem will be established. Some guidelines are emerging.

The pattern-reversal VECP is not a satisfactory test of visual acuity in patients with nystagmus, intraventricular hemorrhages, or shunts. On the other hand, preferential-looking techniques do not always detect strabismic amblyopia[107] and can produce misleading results in patients with oculomotor disorders. A comprehensive pediatric vision-testing laboratory should provide both VECPs and preferential-looking procedures for assessing infant vision.

3-6

PRACTICAL GUIDELINES

3-6-1 Instrumentation

There are a variety of commercial averagers for recording VECPs. The averager used should have an on-line artifact-rejection capability to eliminate unwanted signals from head movements and muscle contractions and a remote-control gating switch to interrupt averaging when the patient stops looking at the stimulus. The pattern stimulator should be checked for luminance contamination. A quick way to do this check is to tape a sheet of white paper over the screen and reverse the checks at an intermediate rate (8–10 reversals/second). If the light seen through the paper is steady, mean luminance is constant. If flicker is seen through the paper, the reversing checks are not matched and the manufacturer should be contacted.

3-6-2 Patient Preparation

Clean a small area of the patient's scalp 1–2 cm above the inion, along the midline, with alcohol or abrasive paste. Tape a gold-cup electrode filled with conducting cream on the area. After cleaning the ears, attach earclip electrodes: the indifferent on one ear, the ground on the other. Figure 3-22 shows a block diagram of a VECP recording system.

3-6-3 Adult Protocol

Use a large field (at least 15°) and two check sizes: small (12–20 minutes) and large (48–60 minutes) (see Figure 3-3). At a distance of 1 meter, a 12-minute check measures 3.5 cm and a 48-minute check measures 14 cm. Testing with small and large checks enables both foveal and perifoveal retinal regions to be tested, thus increasing the diagnostic yield of the test. Patients' pupils should not be dilated (see Figure 3-4). Latency norms should be established for age-matched controls (20 subjects for each decade) starting at age 10. A latency value that is greater than 2.5 standard deviations (99%) should be used as the abnormal cutoff. Absolute amplitude is useful only when comparing interocular differences. Calculate the mean interocular amplitude difference for normal subjects, and use a cutoff of 2.5 standard deviations.

3-6-4 Pediatric Protocol

Since the amplitude and peak latency of specific components change dramatically during the first 6 months of life, it is preferable to record transient rather than

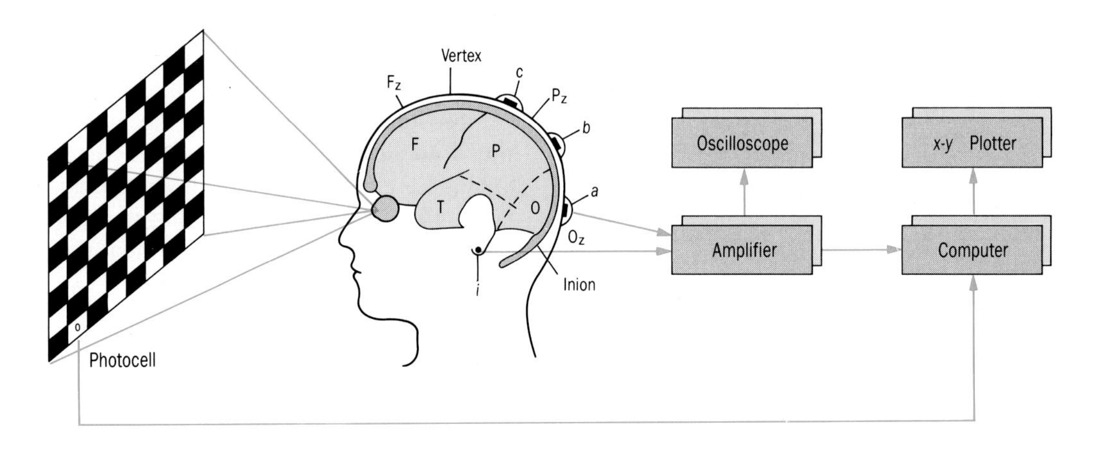

Figure 3-22 *Schematic diagram of electrode locations and electronic equipment for recording the VECP in humans. F,F_z = Frontal; O,O_z = Occipital; P,P_z = Parietal; T = Temporal. The a, b, and c indicate possible locations of active electrodes, and i represents an indifferent electrode at the ear. In this example, the recording configuration is electrode a referenced to the ear (i), which is the input to a differential amplifier. Other recording derivations include a-b, a-c, or b-c (bipolar) or c-i and b-i (referential). The output of the differential amplifier goes to a computer of average transients. The computer is "triggered" by the stimulus, ensuring a time-locked VECP. The oscilloscope can be used to monitor the ongoing VECP activity as well as to provide a view of the final averaged waveform. Finally, a plotter prints out a permanent record of the VECP.*

Reprinted by permission from Sokol S: The visually evoked potential: theory, techniques, and clinical applications. Surv Ophthalmol *1976;21:18–44.*

steady-state VECPs when testing pediatric patients. The patient's fixation must be monitored. An observer can stand behind the video monitor used to generate pattern stimuli and watch for a centered pupillary reflex. The test room should be dark, leaving only the pattern stimulus for the infant to look at. As noted, a remote-control switch, used to start and stop averaging, depending on the patient's fixation, is required, as is an algorithm for rejecting artifacts caused by head movements. Uncooperative patients can be tested while sedated. Wright et al[108] have shown that pediatric patients can be reliably tested under chloral hydrate sedation. These investigators found that measurement of interocular VECP amplitudes was sensitive to the presence of amblyopia in sedated patients.

The optimal check size for infants varies during the first 6 months of life. At 1 month, checks of 120–240 minutes are optimal; at 2 months, 60–120 minutes; at 3–5 months, 30–60 minutes; and at 6 months and older, 15 minutes. If no response is obtained with the check size appropriate for the infant's age, the check size is doubled.

For binocular testing, a check size appropriate for the infant's age is selected first, and larger or smaller checks are used, depending on the outcome (see above). It is often possible to obtain a check-size series and estimate VECP acuity. At least 5 check sizes should be used to fit a reliable curve; for example, 7.5, 15, 30, 60, and 120 minutes. For infants, an intermediate (60-minute) check size is used first. If a signal is obtained, a smaller (30-minute) check size is used; if no signal is obtained, a larger (120-minute) check size is used. If signals are nondetectable with any check size, flash stimuli at 1, 3, 6, 10, and 20 flashes/second are used. Because of the wide variability of the flash VECP, amplitude is not measured. Instead, determine if the response follows the increasing stimulus rate. If it does, it is assumed that the patient has vision of at least light perception.

For monocular testing, patients older than 6 months are first tested binocularly, starting with 15-minute checks. If necessary, check size is increased until a reliable binocular signal is obtained. After a reliable binocular signal has been obtained, each eye is tested, preferably the eye thought to have better acuity first. Since the transient VECP extrapolation technique is too time-consuming for clinical use, an alternative, as described previously, is to measure the difference in amplitude *between* eyes for one check size. Although a visual acuity value is not obtained, there is a greater probability that some clinically useful information will be obtained from each eye.

3-7

SUMMARY

In the measurement of pattern VECP, three parameters can be evaluated: latency, amplitude, and waveform morphology. The first two can be quantified; the third cannot and therefore requires subjective interpretation. The latency of the pattern VECP provides a sensitive means of detecting subclinical lesions of the visual pathways in adults and monitoring developmental changes in infants. The amplitude of the pattern VECP is usually of greater interest in children than is latency because it correlates well with visual acuity, particularly for small checks. Unlike latency, however, it is difficult to use absolute amplitude information because of wide intersubject differences. This drawback can be overcome by using the newer sweep techniques or, when there is a question of monocular visual loss, measuring the relative difference *between* eyes for small checks. While some argue that the VECP provides no more information than does a thorough clinical examination,[109] most agree that the pattern VECP can detect otherwise obscure or clinically inapparent pathology and provide quantitative information unobtainable by even the most astute clinician.[110-116]

REFERENCES

1. Rovamo J, Virsu V: An estimation and application of the human cortical magnification factor. *Exp Brain Res* 1979;**37**:495–510.

2. Daniel PM, Whitteridge W: The representation of the visual field on the cerebral cortex in monkeys. *J Physiol (Lond)* 1961;**159**:203–221.

3. Bodis-Wollner I, Ghialardi MF, Mylin LH: The importance of stimulus selection in VEP practice: the clinical relevance of visual physiology. In: Cracco RQ, Bodis-Wollner I, eds. *Frontiers of Clinical Neuroscience: Evoked Potentials.* New York: AR Liss; 1986:15–27.

4. Riggs LA, Johnson EP, Schick AML: Electrical responses of the human eye to moving stimulus patterns. *Science* 1964;**144**:567.

5. Spekreijse H, van der Tweel LH, Zuidema T: Contrast evoked responses in man. *Vision Res* 1973;**13**:1577–1601.

6. Regan D: *Human Brain Electrophysiology: Evoked Potentials and Evoked Magnetic Fields in Science and Medicine.* New York: Elsevier; 1989.

7. DeVoe RG, Ripps H, Vaughan HC: Cortical responses to stimulation to the human fovea. *Vision Res* 1968; **8**:135–147.

8. Sokol S, Hansen V, Moskowitz A, et al: Evoked potential and preferential looking estimates of visual acuity in pediatric patients. *Ophthalmology* 1983;**90**:552–562.

9. Sokol S, Domar A, Moskowitz A, et al: Pattern evoked potential latency and contrast sensitivity in glaucoma and ocular hypertension. In: Spekreijse H, Apkarian PA, eds. *Electrophysiology and Pathology of the Visual Pathways.* The Hague: Dr W Junk BV Publishers. *Doc Ophthalmol Proc Ser* 1981;**27**:79–86.

10. Sokol S, Moskowitz A, Towle VL: Age-related changes in the latency of the visual evoked potential: influence of check size. *EEG Clin Neurophysiol* 1981;**51**:559–562.

11. Sokol S, Moskowitz A: Effect of retinal blur on the peak latency of the pattern evoked potential. *Vision Res* 1981;**21**:1279–1286.

12. Sokol S: Problems of stimulus control in the measurement of peak latency of pattern visual evoked potential. *Ann NY Acad Sci* 1982;**388**:657–661.

13. Harter MR, White CT: Effects of contour sharpness and check size on visually evoked cortical potentials. *Vision Res* 1968;**8**:701–711.

14. Harter MR: Evoked cortical responses to checkerboard patterns: effects of check size as a function of retinal eccentricity. *Vision Res* 1970;**10**:1365–1376.

15. Seiple WH, Kupersmith MJ, Nelson JI, et al: The assessment of evoked potential contrast thresholds using real-time retrieval. *Invest Ophthalmol Vis Sci* 1984;**25**:627–631.

16. Thompson CR, Harding GF: The visual evoked potential in patients with cataracts. *Doc Ophthalmol Proc Ser* 1978;**15**:193–201.

17. Odom JV, Hobson R, Coldran JT, et al: 10-Hz flash visual evoked potentials predict post-cataract extraction visual acuity. *Doc Ophthalmol* 1987;**66**:291–299.

18. Weinstein GW: Clinical aspects of the visual evoked potentials. *Trans Am Ophthalmol Soc* 1977;**85**:627–673.

19. Fuller DJ, Hutton WL: *Presurgical Evaluation of Eyes with Opaque Media.* New York: Grune & Stratton; 1982.

20. Rinkoff J, de Juan E, Landers M, et al: Flash VEP: useful in management of diabetic vitreous hemorrhage. *Invest Ophthalmol Vis Sci* 1985;**26**(suppl):322.

21. Scherfig E, Edmund J, Tinning S, et al: Flash visual evoked potential as a prognostic factor for vitreous operations in diabetic eyes. *Ophthalmology* 1984;**91**:1475–1479.

22. Papakostopoulos D, Hart CD, Cooper R, et al: Combined electrophysiological assessment of visual system in central serous retinopathy. *EEG Clin Neurophysiol* 1984;**59**:77–80.

23. Folk JC, Thompson HS, Han DP, et al: Visual function abnormalities in central serous retinopathy. *Arch Ophthalmol* 1984;**102**: 1299–1302.

24. Sherman J, Bass SJ, Noble KG, et al: Visual evoked potential delays in central serous choroidopathy. *Invest Ophthalmol Vis Sci* 1986;**27**:214–212.

25. Bass SJ, Sherman J, Bodis-Wollner I, et al: Visual evoked potentials in macular disease. *Invest Ophthalmol Vis Sci* 1985;**26**:1071–1074.

26. Johnson LN, Yee RD, Hepler RS, et al: Alteration of the visual evoked potential by macular holes: comparison with optic neuritis. *Graefes Arch Klin Exp Ophthalmol* 1987;**225**: 123–128.

27. Fukui R, Kato M, Kuroiwa Y: Effect of central scotomata on pattern reversal visual evoked potentials in patients with maculopathy and healthy subjects. *EEG Clin Neurophysiol* 1986;**63**:317–326.

28. Halliday AM, McDonald WI, Mushin J: Delayed visual evoked response in optic neuritis. *Lancet* 1972;**1**:982–985.

29. Hennerici M, Wenzel D, Freund HJ: The comparison of small size rectangle and checkerboard stimulation of the evaluation of delayed visual evoked responses in patients suspected of multiple sclerosis. *Brain* 1977;**100**:119–128.

30. Neima D, Regan D: Pattern VEP and spatial vision in retrobulbar neuritis and multiple sclerosis. *Arch Neurol* 1984;**41**:198–208.

31. Halliday AM, McDonald WI, Mushin J: Visual evoked response in diagnosis of multiple sclerosis. *Br Med J* 1973;**4**:661–664.

32. Asselman P, Chadwick DW, Marsden CD: Visual evoked responses in the diagnosis and management of patients suspected of multiple sclerosis. *Brain* 1975;**98**:261–282.

33. Halliday AM, Halliday E, Kriss A: The pattern-evoked potential in compression of the anterior visual pathways. *Brain* 1976;**99**: 357–374.

34. Kupersmith MJ, Siegel IM, Carr RE, et al: Visual evoked potentials in chiasmal gliomas in four adults. *Arch Neurol* 1981;**38**:362–365.

35. Kramer KK, LaPiana FG, Appleton B: Ocular malingering and hysteria: diagnosis and management. *Surv Ophthalmol* 1979;**24**:89–97.

36. Baumgartner F, Epstein CM: Voluntary alteration of visual evoked potentials. *Ann Neurol* 1982;**12**:475–478.

37. Tan CT, Murray NMR, Sawyers D, et al: Deliberate alteration of the visual evoked potential. *J Neurol Neurosurg Psychiatry* 1984;**47**:518–527.

38. Lentz KE, Chiappa KH: Non-pathologic (voluntary) alteration of pattern-shift visual evoked potentials. *EEG Clin Neurophysiol* 1985;**61**:30P.

39. Chiappa KH, Yiannikas C: Voluntary alteration of evoked potentials. *Ann Neurol* 1982;**12**:496–497.

40. Morgan RK, Nugent B, Harrison JM, et al: Voluntary alteration of pattern visual evoked responses. *Ophthalmology* 1985;**92**:1356–1362.

41. Pritchard WS: Psychophysiology of P300. *Psychol Bull* 1981;**89**:506–525.

42. Towle VL, Sutcliffe E, Sokol S: Diagnosing functional visual deficits with the P300 component of the visual evoked potential. *Arch Ophthalmol* 1985;**103**:47–50.

43. Hoyt CS, Nickel BL, Billson FA: Ophthalmological examination of the infant: developmental aspects. *Surv Ophthalmol* 1982;**26**: 177–189.

44. Abramov I, Gordon J, Hendrickson A, et al: The retina of the newborn human infant. *Science* 1982;**217**:265–267.

45. Yuodelis C, Hendrickson A: A qualitative and quantitative analysis of the human fovea during development. *Vision Res* 1986;**26**: 847–855.

46. Magoun EH, Robb RM: Development of myelin in human optic nerve and tract. *Arch Ophthalmol* 1981;**99**:655–659.

47. Hickey TL: Postnatal development of the human lateral geniculate nucleus: relationship to a critical period for the visual system. *Science* 1977;**198**:836–838.

48. Huttenlocher PR, de Courten C, Garey LJ, et al: Synaptogenesis in human visual cortex: evidence for synapse elimination during normal development. *Neurosci Lett* 1982;**33**: 247–252.

49. Sokol S, Jones K: Implicit time of pattern evoked potentials in infants: an index of maturation of spatial vision. *Vision Res* 1979;**19**: 747–755.

50. Sokol S: Infant visual development: evoked potential estimates. *Ann NY Acad Sci* 1982;**388**:514–525.

51. Moskowitz A, Sokol S: Developmental changes in the human visual system as reflected by the latency of the pattern reversal VEP. *EEG Clin Neurophysiol* 1983;**56**:1–15.

52. Placzek M, Mushin J, Dubowitz LMS: Maturation of the visual evoked response and its correlation with visual acuity in preterm infants. *Dev Med Child Neurol* 1985;**27**: 448–454.

53. Kurtzberg D: Event related potentials in the evaluation of high risk infants. *Ann NY Acad Sci* 1982;**388**:557–571.

54. Mushin J, Dubowitz LMS, Arden GB: Visual function in the newborn infant: behavioral and electrophysiological studies. *Doc Ophthalmol Proc Ser* 1986;**45**:119–135.

55. Ehle A, Sklar FH: Visual evoked potentials in infants with hydrocephalus. *Neurology* 1979;**29**:1541–1544.

56. Sklar FH, Ehle AL, Clark WK: Visual evoked potentials: a non-invasive technique to monitor patients with shunted hydrocephalus. *Neurosurgery* 1979;**4**:529–534.

57. York DH, Pulliam MW, Rosenfeld JG, et al: Relationship between visual evoked potentials and intracranial pressure. *J Neurosurg* 1981;**59**:909–916.

58. Coupland SG, Cochrane DD: Visual evoked potentials, intracranial pressure and ventricular size in hydrocephalus. *Doc Ophthalmol* 1987;**66**:321–330.

59. Guthkelch AN, Sclabassi RJ, Hirsch RP, et al: Visual evoked potentials in hydrocephalus: relationship to head size, shunting and mental development. *Neurosurgery* 1984;**14**:283–286.

60. Sokol S: Measurement of infant visual acuity from pattern reversal evoked potentials. *Vision Res* 1978;**18**:33–39.

61. Regan D: Evoked potentials specific to spatial patterns of luminance and color. *Vision Res* 1973;**13**:2381–2403.

62. Tyler CW, Apkarian P, Levi DM, et al: Rapid assessment of visual function: an electronic sweep technique with pattern visual evoked potential. *Invest Ophthalmol Vis Sci* 1979;**18**:703–713.

63. Norcia AM, Tyler CW: Spatial frequency sweep VEP: visual acuity during the first year of life. *Vision Res* 1985;**25**:1399–1408.

64. Sokol S, Moskowitz A, McCormack G, et al: Infant grating acuity is temporally tuned. *Vision Res* 1988;**28**:1357–1366.

65. Campbell RW, Maffei L: Electrophysiological evidence for the existence of orientation and size detectors in the human visual system. *J Physiol* 1970;**207**:635–652.

66. Regan D: Speedy assessment of visual acuity in amblyopia by the evoked potential method. *Ophthalmologica* 1977;**175**:159–164.

67. Gottlob I, Fendick MG, Guo S, et al: Visual acuity measurements by swept spatial frequency visual-evoked-cortical-potentials (VECPs): clinical applications in children with various visual disorders. *J Pediatr Ophthalmol Strab* 1990;**27**:40–47.

68. Harter MR, Deaton F, Odom JV: Maturation of evoked potentials and visual preference in 6-45 day old infants: effect of check size, visual acuity and refractive error. *EEG Clin Neurophysiol* 1977;**42**:595–607.

69. Harter MR, Suitt CD: Visually evoked cortical responses and pattern vision in the infant: a longitudinal study. *Psychonomic Sci* 1970;**18**:235–237.

70. Sokol S, Dobson V: Pattern reversal visually evoked potentials in infants. *Invest Ophthalmol* 1976;**15**:58–62.

71. DeVries-Khoe LH, Spekreijse H: Maturation of luminance and pattern EPs in man. In: Niemeyer G, Huber CH, eds. *Techniques in Clinical Electrophysiology of Vision.* The Hague: Dr W Junk BV Publishers. *Doc Ophthalmol Proc Ser* 1982;**31**:461-475.

72. Zipf RF: Binocular fixation pattern. *Arch Ophthalmol* 1976;**94**:401–405.

73. Wright KW, Edelman, PM, Walonker F, et al: Reliability of fixation preference testing in diagnosing amblyopia. *Arch Ophthalmol* 1986; **104**:549–553.

74. Sokol S, Bloom B: Visually evoked cortical potentials of amblyopes to a spatially alternating stimulus. *Invest Ophthalmol* 1973;**12**: 936–939.

75. Sokol S, Shaterian ET: The pattern evoked cortical potential in amblyopia as an index of visual function. In: More S, Mein J, Stockbridge L, eds. *Orthoptics: Past, Present and Future.* Miami: Symposia Specialists Medical Books; 1976:59.

76. Sokol S: Abnormal evoked potential latencies in amblyopia. *Br J Ophthalmol* 1983;**67**:310–314.

77. Sokol S: Pattern visually evoked potentials: their use in pediatric ophthalmology. In: Sokol S, ed. *Electrophysiology and Psychophysics: Their Use in Ophthalmic Diagnosis.* International Ophthalmology Clinics. Boston: Little, Brown & Co; 1980;**31**:251–268.

78. Wilcox LM, Sokol S: Changes in the binocular fixation patterns and the visually evoked potential in the treatment of esotropia with amblyopia. *Ophthalmology* 1980;**87**: 1273–1281.

79. Beller R, Hoyt CS, Marg E, et al: Good visual function after neo-natal surgery for congenital monocular cataracts. *Am J Ophthalmol* 1981;**91**:559–565.

80. Gelbart SS, Hoyt CS, Jastrzebski GB, et al: Long term visual results in bilateral congenital cataracts. *Am J Ophthalmol* 1982;**93**: 615–621.

81. Odom JV, Hoyt CS, Marg E: Effect of natural deprivation and unilateral eye patching on visual acuity of infants and children: evoked potential measurements. *Arch Ophthalmol* 1981;**99**:1412–1416.

82. Hoyt CS, Jastrzebski GB, Marg E: Amblyopia and congenital esotropia: visually evoked potential measurement. *Arch Ophthalmol* 1984; **102**:58–61.

83. Millodot M, Riggs LA: Refraction determined electrophysiologically: responses to alternation of visual contours. *Arch Ophthalmol* 1970;**84**:272–278.

84. Regan D: Rapid objective refraction using evoked brain potentials. *Invest Ophthalmol* 1973;**19**:669–679.

85. Cogan DG: Congenital ocular motor apraxia. *Can J Ophthalmol* 1966;**4**:253–259.

86. Gittinger JW Jr, Sokol S: The visual-evoked potential in the diagnosis of congenital ocular motor apraxia. *Am J Ophthalmol* 1982;**93**:700–703.

87. Duchowny MD, Weiss P, Heshmatolah M, et al: Visual evoked responses in childhood cortical blindness after head trauma and meningitis. *Neurology* 1974;**24**:933–940.

88. Frank Y, Torres F: Visual evoked potentials in the evaluation of "cortical blindness" in children. *Ann Neurol* 1979;**6**:126–129.

89. Weiskrantz L, Warrington EK, Sanders MD, et al: Visual capacity in the hemianopic field following a restricted occipital ablation. *Brain* 1974;**97**:708–728.

90. Jan JE, Farrel K, Wong PK, et al: Eye and head movements of visually impaired children. *Dev Med Child Neurol* 1986;**28**:285–293.

91. Bodis-Wollner I, Atkin A, Raab E, et al: Visual association cortex in vision in man: pattern evoked potentials in a blind boy. *Science* 1977;**198**:639–640.

92. Regan D, Regal DM, Tibbles JAR: Evoked potentials during recovery from blindness recorded serially from an infant and his normally sighted twin. *EEG Clin Neurophysiol* 1982;**54**:465–468.

93. Kupersmith MJ, Nelson JI: Preserved visual evoked potential in infant cortical blindness. *Neuro-ophthalmology* 1986;**6**:85–94.

94. Sokol S, Hedges TR Jr, Moskowitz A: Pattern VEPs and preferential looking acuity in infantile traumatic blindness. *Clin Vision Sci* 1987;**2**:59–61.

95. Illingworth, RS: Delayed visual maturation. *Arch Dis Child* 1961;**36**:407–409.

96. Mellor DH, Fielder AR: Dissociated visual development: electrodiagnostic studies in infants who are "slow to see." *Dev Med Child Neurol* 1980;**22**:327–335.

97. Uemura Y, Oguchi Y, Katsumi O: Visual developmental delay. *Ophthalmic Pediatr Genet* 1981;**1**:49–58.

98. Watanabe K, Iwase K, Hara K: Maturation of visual evoked responses in low birth weight infants. *Dev Med Child Neurol* 1972;**14**:425–435.

99. Hoyt CS, Jastrzebski GB, Marg E: Delayed visual maturation in infancy. *Br J Ophthalmol* 1983;**67**:127–130.

100. Lambert SR, Kriss A, Taylor D: Delayed visual maturation: a longitudinal clinical and electrophysiological assessment. *Ophthalmology* 1989; **96**: 524–529.

101. Mantyjarvi MI: The amblyopic school girl syndrome. *J Pediatr Ophthalmol Strab* 1981;**18**:30–33.

102. van Balen TM, Slijper FEM: Psychogenic amblyopia in children. *J Pediatr Ophthalmol Strab* 1978;**15**:164–167.

103. Skarf B, Panton C: VEP testing in neurologically impaired and developmentally delayed infants and young children. *Invest Ophthalmol Vis Sci* 1987;**28**(suppl):302.

104. Mohn G, van Hoff-van Duin J: Behavioral and electrophysiological measures of visual functions in children with neurological disorders. *Behav Brain Res* 1983;**10**:177–187.

105. Creel D, Spekreijse H, Reits D: Evoked potentials in albinos: efficacy of pattern stimulus in detecting misrouted optic fibers. *EEG Clin Neurophysiol* 1981;**52**:595–603.

106. Teller D, McDonald M, Preston K, et al: Assessment of visual acuity in infants and children. *Dev Med Child Neurol* 1986;**28**:779–789.

107. Mayer DL, Fulton AB, Rodier D: Grating and recognition acuities of pediatric patients. *Ophthalmology* 1984;**91**:947–953.

108. Wright KW, Eriksen J, Shors TJ, et al: Recording pattern visual evoked potentials under chloral hydrate sedation. *Arch Ophthalmol* 1986;**104**:718–721.

109. Burde RM: Discussion: voluntary alternation of pattern visually evoked responses. *Ophthalmology* 1985;**92**:1362–1363.

110. Isen A, Cracco RQ: Overuse of evoked potentials: caution. *Neurology* 1983;**33**:618–621.

111. Smith DN: The clinical usefulness of the visual evoked response, *J Pediatr Ophthalmol Strab* 1984;**21**:235–236.

112. Hoyt CS: The clinical usefulness of the visual evoked response. *J Pediatr Ophthalmol Strab* 1984;**21**:231–234.

113. Cracco RQ: Utility of evoked potentials. *J Am Med Assn* 1985;**254**:3490.

114. Bodis-Wollner I: Utility of evoked potentials. *J Am Med Assn* 1985;**254**:3490.

115. Kimura J: Abuse and misuse of evoked potentials as a diagnostic test. *Arch Neurol* 1985;**42**:78–80.

116. Spekreijse H, Apkarian P: The use of a system analysis approach to electrodiagnostic (ERG and VEP) assessment. *Vision Res* 1986; **26**:195–219.

117. Atkinson J, Braddick OJ, Moar K: Development of contrast sensitivity over the first 3 months of life in the human infant. *Vision Res* 1977;**17**:1037–1044.

118. Harris L, Atkinson J, Braddick O: Visual contrast sensitivity in a 6 month old infant measured by the evoked potential. *Nature* 1976;**264**:570–571.

119. Marg E, Freeman RD, Peltzman P: Visual acuity and development in human infants: evoked potential measurements. *Invest Ophthalmol* 1976;**15**:150–152.

120. Pirchio M, Spinelli D, Fiorentini A, et al: Infant contrast sensitivity evaluated by evoked potentials. *Brain Res* 1978;**141**:179–184.

121. Sokol S, Moskowitz A: Comparison of pattern VEPs and preferential looking behavior in 3-month-old infants. *Invest Ophthalmol Vis Sci* 1985;**126**:359–365.

Supplemental Readings

Bodis-Wollner I, ed. *Evoked Potentials.* New York: New York Academy of Sciences; 1982: vol 388.

Birch EE: Visual acuity testing in infants and young children. *Ophthalmol Clin North Am* 1989;**2**:369–389.

Chiappa KH: *Evoked Potentials in Clinical Medicine.* New York: Raven Press; 1983.

Cracco RQ, Bodis-Wollner I, eds. *Frontiers of Clinical Neuroscience: Evoked Potentials.* New York: AR Liss; 1986.

Fulton AB, Hartmann EE, Hansen RM: Electrophysiologic testing techniques for children. *Doc Ophthalmol* 1989;**71**:341–354.

Regan D: *Human Brain Electrophysiology: Evoked Potentials and Evoked Magnetic Fields in Science and Medicine.* New York: Elsevier; 1989.

Sokol S: Pattern visually evoked potentials: their use in pediatric ophthalmology. In: Sokol S, ed. *Electrophysiology and Psychophysics: Their Use in Ophthalmic Diagnosis.* International Ophthalmology Clinics. Boston: Little, Brown & Co; 1980;**31**:251–268.

Skarf B: Clinical use of visual evoked potentials. *Ophthalmol Clin North Am* 1989;**2**: 499–518.

Spehlmann R: *Evoked Potential Primer: Visual, Auditory Somatosensory Evoked Potentials in Clinical Diagnosis.* Boston: Butterworth; 1985.

Name ____

Address ____

City and State ___________________________________ Zip ________

Telephone (________) ______________ *Academy Member ID# ______________
area code

**Your ID Number is located following your name on any Academy mailing label, in your Membership Directory, and on your Monthly Statement of Account.*

CATEGORY 1 CME CREDIT FORM

Ophthalmology Monographs 2

Electrophysiologic Testing in Disorders of the Retina, Optic Nerve, and Visual Pathway

You may claim 1 hour of Category 1 Continuing Education Credit, up to a 25-hour maximum for each hour you spend studying this Ophthalmology Monograph. If you wish to claim continuing education credit for your study of this monograph, you must complete and return the self-study examination answer sheet on the back of this page, along with the following signed statement, to the Academy offices:

American Academy of Ophthalmology

P.O. Box 7424

San Francisco, CA 94120-7424

ATTN: Education Department

I hereby certify that I have spent ______ (up to 25) hours of study on the Ophthalmology Monograph *Electrophysiologic Testing in Disorders of the Retina, Optic Nerve, and Visual Pathway* and that I have completed the self-study examination. (The Academy *upon request* will send you a transcript of the credits listed on this form. You can check the box below if you wish credit verification now.)

☐ Please send credit verification now.

Signature ________________________________ ______________
Date

MONOGRAPH COMPLETION FORM

Ophthalmology Monographs 2

Answer Sheet for *Electrophysiologic Testing in Disorders of the Retina, Optic Nerve, and Visual Pathway*

Question	Answer				
1	a	b	c	d	e
2	a	b	c	d	e
3	a	b	c	d	e
4	a	b	c	d	e
5	a	b	c	d	e
6	a	b	c	d	e
7	a	b	c	d	e
8	a	b	c	d	e
9	a	b	c	d	e
10	a	b	c	d	e
11	a	b	c	d	e
12	a	b	c	d	e
13	a	b	c	d	e
14	a	b	c	d	e
15	a	b	c	d	e
16	a	b	c	d	e
17	a	b	c	d	e
18	a	b	c	d	e
19	a	b	c	d	e
20	a	b	c	d	e

SELF-STUDY EXAMINATION

The self-study examination for *Electrophysiologic Testing in Disorders of the Retina, Optic Nerve, and Visual Pathway* consists of 20 multiple-choice questions and is intended for use *following* completion of the monograph. Questions are constructed so that there is one "best" answer. For each question, record your initial impression on the answer sheet by circling the appropriate letter. It is recognized that a disagreement about the optimal answer may occur despite the attempt to avoid ambiguous selections. A discussion of the most appropriate answer to each question follows the examination. Answers should not be consulted until the entire examination has been completed.

QUESTIONS

1. Reduction in ERG oscillatory potential amplitudes is most likely to occur in all of the following disorders *except:*

a. open-angle glaucoma

b. X-linked juvenile retinoschisis

c. diabetic retinopathy

d. congenital stationary night blindness

e. central retinal vein occlusion

2. Which of the following tests is most likely to be abnormal in a patient with chronic open-angle glaucoma?

a. early receptor potential

b. focal, foveal cone ERG

c. full-field ERG

d. pattern-evoked ERG

e. EOG

3. In which of the following disorders is a selective or predominant reduction in the ERG b-wave *not* found?

a. congenital stationary night blindness

b. Oguchi's disease

c. X-linked juvenile retinoschisis

d. ophthalmic artery occlusion

e. quinine intoxication

4. In which of the following macular dystrophies is the ERG most likely to be reduced in amplitude?

a. Stargardt's macular dystrophy

b. Best's (vitelliform) macular dystrophy

c. butterfly dystrophy

d. central areolar choroidal atrophy

e. X-linked juvenile retinoschisis (clinically limited to the fovea)

5. In which of the following inflammatory disorders is the ERG most likely to be reduced in amplitude compared to a normal population?

a. diffuse rubella retinopathy

b. toxoplasmosis

c. multiple evanescent white dot syndrome

d. histoplasmic choroiditis

e. luetic chorioretinitis

6. In which of the following disorders is a reduction in ERG amplitude most likely to be seen?

a. chloroquine or hydroxychloroquine retinopathy

b. chlorpromazine retinopathy

c. thioridazine chorioretinopathy

d. indomethacin maculopathy

e. rubella retinopathy

7. In oculocutaneous albinism, the EOG light-peak to dark-trough ratio response is best categorized as:

a. subnormal

b. markedly subnormal

c. supernormal

d. normal

e. less than 1.00

8. In which of the following is the EOG light-peak to dark-trough ratio most likely to be abnormal?

a. diffuse rubella retinopathy

b. diffuse cone dystrophy

c. Stargardt's macular dystrophy

d. familial drusen of Bruch's membrane

e. central retinal artery occlusion

9. In which of the following disorders is the EOG light-peak to dark-trough ratio *most* likely to be abnormal?

a. X-linked recessive congenital stationary night blindness

b. autosomal recessive congenital stationary night blindness

c. Oguchi's disease

d. chloroquine retinopathy

e. gyrate atrophy of the choroid and retina

10. The most characteristic ERG finding seen in patients with a retained intraocular iron foreign body includes:

a. selective or predominant reduction in b-wave amplitude

b. nondetectable response

c. supernormal a-wave

d. supernormal b-wave under photopic conditions

e. selective loss of oscillatory potentials

11. The light-sensitive component, or slow light rise, of the EOG is generated by which of the following?

a. depolarization of the basal membrane of the retinal pigment epithelial cells

b. hyperpolarization of the basal membrane of the retinal pigment epithelial cells

c. depolarization of retinal photoreceptor cells

d. exclusively cone photoreceptor cells

e. primarily retinal pigment epithelial cells within the fovea

12. Which of the following statements regarding the EOG is false?

a. The EOG light-peak to dark-trough ratio is generally normal in patients with Stargardt's macular dystrophy.

b. The light rise of the EOG positively correlates with the amount of melanin within the retinal pigment epithelial cells.

c. The EOG response is influenced by diffuse disease of rod photoreceptor cells.

d. The EOG ratio is likely to be normal in carriers of choroideremia.

e. The EOG ratio is abnormal in patients with Best's macular dystrophy even when retinal lesions are not apparent clinically.

13. Cone function can best be isolated from rod function on ERG testing by using:

a. a 10-Hz flickering white stimulus after dark adaptation

b. a single-flash, high-intensity white light after 30 minutes of dark adaptation

c. a 10-Hz flickering blue light stimulus after dark adaptation

d. a high-intensity, white light, single-flash stimulus in the presence of an adapting background

e. a 10-Hz flickering red stimulus after dark adaptation

14. The EOG is most likely to be reduced in which of the following?

a. primary optic atrophy

b. glaucoma

c. carotid artery occlusion

d. optic disc drusen

e. atrophic foveal lesions in association with congenital toxoplasmosis

15. Which of the following is least likely to be observed on ERG testing?

a. normal a- and b-wave amplitudes but a nondetectable c-wave

b. reduced a-wave, b-wave, and oscillatory potentials

c. reduced a-wave and normal b-wave

d. reduced b-wave and oscillatory potentials but normal a-wave

e. reduced ERG amplitudes in a *carrier* of X-linked retinitis pigmentosa

16. The pattern VECP in a patient with a history of optic neuritis will have:

a. a normal amplitude and a prolonged latency

b. an abnormal amplitude and a prolonged latency

c. an abnormal amplitude and a normal latency

d. a normal amplitude and a normal latency

17. In which condition are the amplitude and the latency of the pattern VECP for small checks likely to be normal?

a. amblyopia

b. retinitis pigmentosa

c. optic neuritis

d. uncorrected myopia

e. congenital nystagmus

18. Which of the following statements is false?

a. Pattern VECPs can be normal in a blind patient.

b. Pattern VECPs are abnormal in an uncorrected high myope.

c. Flash VECPs are abnormal in patients with elevated intracranial pressure.

d. The pattern VECP is normal in central serous retinopathy.

19. Which of the following tests would most likely detect amblyopia?

a. flash VECP recorded with a 30-flashes/second test stimulus

b. pattern VECP recorded with 10 minutes of arc checks

c. pattern VECP recorded with 60 minutes of arc checks

d. flash ERG recorded with 30-flashes/second test stimulus

20. To evaluate an infant thought to be "blind," which combination of tests would be most useful?

a. flash ERG and OKN testing

b. pattern VECP and flash ERG

c. pattern VECP and preferential-looking acuity test

d. pattern and flash VECPs

ANSWERS AND DISCUSSION

These answers and explanations are to help you confirm that the reasoning you used in finding the most appropriate answer was correct. If you missed the question, the answer may help you to decide whether it was due to misinterpretation of the question or to poor wording. If, instead, you missed the question because of miscalculation or failure to recall relevant information, the answer and the explanation may help fix the principle in your memory.

1. **Answer–a.** Circulatory disturbances, such as those occurring in patients with diabetic retinopathy and central retinal vein occlusion, are known to reduce the oscillatory potential amplitudes. Patients with X-linked juvenile retinoschisis and congenital stationary night blindness can also manifest reduction in these potentials. As with the a- and b-waves, the generators of the oscillatory potentials are not affected in patients with chronic open-angle glaucoma.

2. **Answer–d.** The early receptor potential, focal, and full-field ERG are predictably abnormal in disorders of the photoreceptor cells. The EOG is not generated by either the ganglion cells or the optic nerve fibers. Currently available information would support the conclusion that ganglion cells are likely to be primary generators of a pattern-evoked ERG.

3. **Answer–d.** Since the ophthalmic artery provides circulation to the photoreceptor cells, an occlusion of this vessel would predictably affect the a- as well as the b-wave amplitude. The other four disorders involve structures in the more proximal portion of the retina where the b-wave is generated.

4. **Answer–e.** Patients with X-linked juvenile retinoschisis, even with retinal changes clinically limited to the fovea, will show a selective reduction in the b-wave amplitude. The other four disorders, in the majority of instances, are associated with normal ERG a- and b-wave amplitudes.

5. **Answer–c.** Somewhat atypical for presumed inflammatory lesions, patients with the multiple evanescent white dot syndrome can have appreciable reduction in ERG amplitudes. Of interest: as the disorder remits, the ERG amplitudes improve and can return to normal. The other four disorders either do not affect photoreceptor cells or affect these structures in a more focal manner that will not reduce ERG amplitudes unless diffuse regions of the photoreceptor cells are impaired.

6. **Answer–c.** The use of chlorpromazine, chloroquine, and indomethacin either is not likely to affect retinal structures or, if degenerative changes in the retina become apparent, is more likely to become focal in nature and not affect the full-field ERG. Patients with thioridazine chorioretinopathy are more likely to manifest diffuse impairment of photoreceptor cell function detectable by ERG measurements.

7. **Answer–d.** Since retinal pigment epithelial cell function is otherwise normal, hypopigmentation of the fundus in patients with oculocutaneous albinism is not associated with any reduction in the EOG light-peak to dark-trough ratio.

8. **Answer–e.** Patients with central retinal artery occlusion are likely to have an abnormality in the EOG light rise. Thus, the EOG light-peak to dark-trough ratio tends to be reduced in such patients. Patients with the other four disorders tend not to manifest abnormal EOG ratios unless they have markedly extensive atrophy throughout the retina, which can be seen in the advanced stages of Stargardt's macular dystrophy (fundus flavimaculatus).

9. **Answer–e.** Patients with gyrate atrophy of the choroid and retina tend to have diffuse retinal pigment epithelial and photoreceptor cell disease. These changes can be ascertained by reduction in ERG amplitudes as well as in EOG light-peak to dark-trough ratios. The other four disorders would not tend to affect structures within the retina responsible for generating the EOG light rise.

10. **Answer–a.** A selective or predominant reduction in b-wave amplitude is characteristically reported at an intermediate stage of retinal toxicity in patients with a retained intraocular iron foreign body. Although a nondetectable response could result in very advanced stages, the more characteristic finding is that of a selective or predominant b-wave reduction. A supernormal response involving primarily the b-wave has been reported in at least some patients. However, this finding is seen most frequently under scotopic conditions and overall is less frequently found than is a selective or predominant reduction in the b-wave amplitude.

11. **Answer–a.** The light-sensitive component of the EOG is generated by a depolarization of the basal membrane of the retinal pigment epithelial cells. Cone photoreceptor cells are not exclusively involved in the generation of the light rise. In vertebrates, photoreceptor cells hyperpolarize rather than depolarize and thus their depolarization would not relate to the EOG light rise. Since the EOG is a diffuse response, retinal pigment epithelial cells within the fovea would not provide a major contribution to the light rise.

12. **Answer–b.** Although the EOG light rise does relate to retinal pigment epithelial and photoreceptor cell function, the amount of melanin pigment within retinal pigment epithelial cells does not influence the EOG light-peak to dark-trough ratio in any known fashion or to any appreciable degree. The other four statements are true.

13. **Answer–d.** A 10-Hz stimulus, particularly when applied under dark-adapted conditions, will not isolate cone function. Similarly, a single-flash, high-intensity white light stimulus during dark adaptation consists of a combined rod and cone response that is rod-dominant. Cone function can best be assessed in the presence of an adapting background light and with the use of a high-intensity stimulus.

14. **Answer–c.** Disorders of the optic nerve or ganglion cells do not traditionally affect EOG light-peak to dark-trough ratios. Similarly, a local inflammatory lesion within the fovea will not affect the EOG, which represents a diffuse response from the retina. However, extensive carotid artery occlusion, with impairment of retinal circulation, will affect the EOG because of ischemia to both the outer and the inner retina.

15. **Answer–c.** Since the generation of the ERG b-wave depends on the photoreceptor cell changes that generate the a-wave, a reduction in the ERG a-wave will have an obligatory effect on the b-wave amplitude.

16. **Answer–a.** The important point here is a history of optic neuritis. During the acute stage of optic neuritis, VECP amplitude is reduced, compatible with the patient's poor acuity, and P_1 latency is prolonged. After the patient's recovery, amplitude is normal while latency continues to be abnormal.

17. **Answer–b.** Since the pattern VECP is primarily a reflection of the central visual field, patients with retinitis pigmentosa will have normal pattern VECPs, unless their central vision is impaired.

18. Answer–d. The amplitude *and* the latency of the pattern VECP are abnormal in some retinal diseases. However, with recovery of visual function, both amplitude and latency are normal—unlike optic neuritis, where the latency abnormality persists.

19. Answer–b. VECPs recorded with small checks (<20 minutes) are most sensitive to reduced acuity. If VECPs are recorded with large checks, reduced acuity will go undetected. Neither flash ERGs nor flash VECPs are sensitive to acuity loss; a patient with counting-fingers vision can still have a normal flash VECP.

20. Answer–b. Flash ERGs should be recorded from any infant thought to be blind to rule out the possibility of retinal disease. Three retinal diseases that can present with a relatively normal fundus and an abnormal ERG (and VECP) in a young infant include Leber's congenital amaurosis, congenital stationary night blindness, and rod monochromatism.

INDEX

NOTE: A *t* following a page number indicates tabular material, and an *f* following a page number indicates an illustration. Drugs are listed under their generic names; when a drug trade name is listed, the reader is referred to the generic name.

T

U

V